African Cichlids I
Malawi
Mbuna

Erwin Schraml

Die Deutsche Bibliothek - CIP-Einheitsaufnahme

Aqualog : reference fish of the world. -
Mörfelden-Walldorf : A.C.S.
African Cichlids. I (Malawi, Mbuna).-1998

African Cichlids / Erwin Schraml. - Mörfelden-Walldorf : A.C.S. (Aqualog)

NE: Schraml, E. 1 (1998)

ISBN 3-931702-79-0

Rothwiesenring 5, 64546 Mörfelden-Walldorf / Germany

Text und fachliche Bearbeitung
Erwin Schraml
Übersetzungen
Deutsch-Englisch: Mary Bailey
Englisch-Deutsch: Frank Schäfer
Organisation
Wolfgang Glaser
Redaktion
Frank Schäfer
Titelgestaltung:
Gaby Geiß, Büro für Grafik, Frankfurt a.M.
Druck, Satz, Verarbeitung:
Lithos Verlag A.C.S.
Bildbearbeitung Frank Teigler
Giese Druck, Offenbach
Gedruckt auf EURO ART glänzend,
100% chlorfrei von PWA Umweltfreundlich

Printed in Germany

Inhalt
Contents

* by Ad Konings ** by Karen A. Kellog & Jay R. Stauffer, jr.

Erklärungen der Abkürzungen in den wissenschaftlichen Namen
Key to the abbreviations of the scientific names

Beispiel/ *example:*	*Pseudotropheus*	*tropheops*	*tropheops*	REGAN,	1922
	Gattung *Genus*	Art *Species*	Unterart *Subspecies*	Beschreiber *Describer*	Publikationsjahr *Year of publication*

sp./spec.: = species (lat.): Art / *species*
Hinter einem Gattungsnamen meint dies: Ein Artname steht (noch) nicht zur Verfügung, die Art ist bislang nicht eindeutig bestimmt bzw. noch nicht formell beschrieben./
Following the genus name, this means: A species name is not yet available, the species has not yet been determined or formally described.

ssp.: = subspecies (lat.): Unterart / *subspecies*
Einige Arten haben ein sehr großes Verbreitungsgebiet; innerhalb dieses Gebietes gibt es Populationen, die sich äußerlich zwar deutlich von anderen Populationen unterscheiden, genetisch jedoch zur gleichen Art gehören. Solche Populationen erhalten als Unterart einen wissenschaftlichen Namen. Ist die Unterart bislang unbenannt, so steht hier nur ssp./
Some species inhabit an area of a very wide range; within this area, there may be populations which differ significantly in appearance from other populations, but clearly belong to the same species. Such populations may get a third scientific name, indicating a subspecies. If a subspecies name has not yet been formally given, the abbreviation ssp. is added.

cf.: = confer (lat.): vergleiche / *compare*
Einem Artnamen vorangestellt meint dies: Das vorliegende Exemplar oder die entsprechende Population weicht in gewissen Details von der typischen Form ab, jedoch nicht so gravierend, daß es oder sie einer anderen Art zugeordnet werden könnte./
Placed in front of a species name this means: The specimen shown or the respective population to which it belongs differ in some minor details from the typical form, but these differences do not justify to place it into a species of its own .

sp. aff.: = species affinis (lat.): ... ähnliche Art / *similar to ...*
Einem Artnamen vorangestellt meint dies: Die vorliegende Art ist bisher noch nicht bestimmt, sie ähnelt jedoch der genannten und bereits beschriebenen Art./
Placed in front of a species name, this means: The species at hand is not yet determined but it is very similar to the one named in the following.

Hybride/*hybrid*: Kreuzungsprodukt, Mischling zweier Arten / *hybrid or crossbreed of two species*

leg. : = legat (lat.): Überbringer, Bote / *a person who brings or gives something*
Im AQUALOG - Kontext bedeutet dies: die exakte Fundortangabe und/oder die Determination zu dem abgebildeten Fisch stammt von der in Kapitälchen geschriebenen Person / *This means in the AQUALOG context: the information about the exact place of origin and/or the determination of the fish depicted is given by the person written in capitals*

Mbuna - was ist das?

Schon seit ihrer Einführung in die Aquaristik sind die in diesem Band vorgestellten Fischarten als "Mbunas" bekannt. Der Name ist der Chitonga-Sprache entlehnt und wird von den am Malawisee heimischen Fischern für die in der Felsküste vorkommenden Cichliden verwendet. In der wissenschaftlichen Literatur hat sich der Begriff ebenfalls durchgesetzt, weil er auch als Sammelbezeichnung für eine gut unterscheidbare Gruppe miteinander verwandter Arten taugt. Mbunas waren auch die ersten Arten, die für den kommerziellen Zierfischhandel aus dem Malawisee exportiert wurden. Das früheste Datum (1961) das für den Beginn der Ausfuhr dieser Fische zu recherchieren war, bezieht sich auf nur wenige Ausstellungsexemplare. Ein umfangreicherer Handel begann erst Mitte der 60er Jahre und selbst Anfang der 70er Jahre waren nur wenige Arten bei den Aquarianern wirklich bekannt. *Melanochromis auratus, Labeotropheus fuelleborni* und *L. trewavasae*, sowie Arten des *Pseudotropheus zebra*-(jetzt *Maylandia*) und *P. tropheops*-Komplexes waren die Ersten. Die damals entstandenen Cichliden-Vereinigungen trugen rasch zu einer größeren Verbreitung dieser Fische bei. Weil der Malawisee nie verkehrsgünstige Anbindung an Seehäfen hatte, gab es vor den Zeiten des Luftfrachtversandes auch keine frühen Importe von Fischen aus diesem Gebiet, wie es sonst bei vielen anderen tropischen Arten der Fall war.

Der wahrscheinlich erste Artikel über Malawisee-Cichliden in einer Aquarienzeitschrift stammt aus dem Jahr 1939 (erschienen in "Water Life") und war von E. TREWAVAS verfaßt, die sich schon 1935 in einer wissenschaftlichen Pionierarbeit mit diesen Fischen befaßt hatte. In Deutschland erschien 1965 der erste Bericht in einer Aquarienzeitschrift, aber erst Anfang der 70er Jahre folgten weitere, die bald in kürzeren Abständen immer wieder "neue" Arten den interessierten Aquarianern vorstellten. Überhaupt war das Vorhandensein einer schier endlos scheinenden Formenfülle einer der Mechanismen, die das Interesse an diesen Fischen nie erlahmen lies. Auch heute ist noch mit jeder Menge weiterer Formen oder gar Arten zu rechnen, weil der riesige Malawisee bei weitem noch nicht an allen Ecken betaucht wurde.

Der Malawisee ist heute eines der interessantesten Reiseziele für fernreisende Aquarianer. Und so ist es nicht verwunderlich, daß neben den Exporteuren eine Vielzahl von tauchenden und schnorchelnden Aquarianern immer wieder neue Arten oder Formen entdecken. Diese Fische werden häufig auch für die Aquaristik exportiert und natürlich gibt es bisher noch immer nicht für alle Formen gültige wissenschaftliche Namen. Nicht nur in der Aquaristik behilft man sich mit Händlerbezeichnungen; in wissenschaftlichen Kreisen wird das lediglich anders genannt (Cheironym oder Arbeitsbezeichnung). Da es oft Jahre oder Jahrzehnte dauert, bis eine bekannte Art wissenschaftlich benannt wird, spielen diese vorläufigen Namen eine wichtige Rolle, wenn es um die Kommunikation über Mbunas geht. Im Index sind deshalb viele Arbeitsbezeichnungen

Mbuna - what are they?

Ever since their original introduction to the aquarium hobby the fish species covered by this volume have been known as "mbuna". The name is borrowed from the Chitonga language and is used by the native fishermen around Lake Malawi to describe the cichlids found along the rocky coastline. The name has also found its way into the scientific literature, where it has been applied as a collective term for a well-defined group of related species. Mbuna were also the first species to be exported from Lake Malawi for the commercial ornamental fish trade. The earliest date (1961) revealed by research applies only to a few specimens for display purposes. It was not until the middle 1960s that more extensive trade began, and even by the beginning of the 1970s only a few species were actually known to aquarists. *Melanochromis auratus, Labeotropheus fuelleborni* and *L. trewavasae*, as well as members of the *Pseudotropheus zebra* (now *Maylandia*) and *Ps. tropheops* complexes, were the first. A number of cichlid associations were founded round about this time, and quickly led to a wider distribution of these fishes. Because there has never been any practicable connection, transport-wise, between Lake Malawi and any seaport, before the days of air freight there was no possibility of earlier imports of fishes from this area, as it was the case with many other tropical species.

Probably the first article on mbuna in an aquarium magazine dates from 1939 (in "Water Life"), and was written by Dr. Ethelwynn TREWAVAS, who had earlier (1935) published a pioneering scientific work on these fishes. The first article in a German aquarium magazine appeared in 1965, but with no further such items until the beginning of the 70s. Before long such articles started to appear in rapid succession, informing interested aquarists about the steady influx of "new" species. It was undoubtedly the existence of an apparently endless host of varieties that served to prevent interest in these fishes from waning. Even today there are still numbers of new forms, and even species, to be expected, as there is still a long way to go before all the corners of the huge lake will have been thoroughly investigated.

Today Lake Malawi has become an interesting destination for aquarists with the urge to visit distant places. And thus is is hardly surprising that not only the exporters, but also a large number of diving and snorkelling aquarists are continually discovering new species. These fishes are often subsequently exported for the aquarium hobby, and not surprisingly there are not yet valid scientific names for all the different forms.
This gap has been filled by the use of "trade names"; temporary names are also used in scientific circles (cheironyms or working names). Because it may take years or even decades before a known species is scientifically named, these temporary names play an important role as regards communication concerning mbuna. For this reason the index in this book includes many working names which in turn refer to older designations, as some species have been given such names several times.

aufgeführt, die auf ältere Bezeichnungen verweisen, weil manche Arten mehrmals "benamt" wurden.

Mbunas in der Wissenschaft

Mbunas sind Buntbarsche, die verwandtschaftlich ebenso zu den haplochrominen Arten zu rechnen sind, wie der überwiegende Teil der anderen Cichliden des Malawisees. Fast alle Arten dieses Komplexes sind schon alleine dadurch charakterisiert, daß sie streng substratgebunden an Felsküsten leben. Sie haben aber auch morphologische Kennzeichen gemeinsam, die aufgrund ihrer Lebensweise entstanden sind. So haben alle Arten im Brustbereich viele kleine, tief eingebettete Schuppen, eine Anpassung an das scharfkantige Substrat, das zu locker angewachsene Schuppen bei Berührung leicht abreißen könnte. Im seitlichen Verlauf werden sie abrupt von größeren Schuppen abgelöst, die dann die Flanken bedecken. Die meisten Mbunas besitzen echte Eiflecken in der Afterflosse (nicht alle, so fehlt z.B. der überwiegenden Mehrzahl der Weibchen von *Melanochromis auratus* dieses Merkmal), die bei Männchen von einem Hof umgeben sind. Eine anatomische Besonderheit ist auch die Reduktion des linken Eierstocks.

Je nach Artkonzept ist die Rede von bis zu 1.600 Cichlidenarten, die im Malawisee heimisch sein sollen. Es dürfte daher jedem einleuchten, daß es für einen einzelnen Wissenschaftler nicht möglich ist, eine komplette Neugruppierung von allen Arten vorzunehmen. Deshalb muß jede Arbeit, die einen Teil dieser Fische berücksichtigt, wie Stückwerk wirken und wird fast zwangsläufig auch Fehler enthalten. Dennoch müssen wir mit diesem Manko leben, weil nur so Diskussionsgrundlagen für weitere Arbeiten geschaffen werden können.

Die Gattungen (in chronologischer Reihenfolge)

Gephyrochromis Boulenger, 1901 war die erste eigenständig beschriebene Gattung von Mbuna-Cichliden. Als Besonderheit gilt die häufig konische Ausprägung der Zähne der äußeren Zahnreihen in beiden Kiefern sowie die Tatsache, daß die einzelnen Zähne länger und schlanker sind als bei der nahverwandten Gattung *Pseudotropheus.* Ansonsten finden sich viele Übereinstimmungen in beiden Gattungen. Aquaristisch ist wenig über *Gephyrochromis* bekannt geworden auch wenn gelegentliche Einfuhren die Fische nicht völlig unbekannt erscheinen lassen. Im See werden sie einzeln aber auch in größeren Gruppen angetroffen.

Cynotilapia und *Pseudotropheus* wurden beide von Regan, 1922 beschrieben. Beide mit Typusarten die von Günther bereits 1864 in den damaligen Cichliden-Sammelgattungen *Hemichromis* bzw. *Chromis* erstbeschrieben waren.

Die Vertreter der Gattung *Cynotilapia* besitzen in der äußeren Reihe lange, mit großem Abstand zueinander stehende, einspitzige, stark konische Zähne. Die inneren Zähne sind kleiner und haben unregelmäßige Abstände zueinander. Im Aquarium wie auch in der Natur sind die Männchen stark

Mbuna in Science

Mbuna are cichlids, and are, phylogenetically speaking, included among the haplochromine species, as are the vast majority of other Lake Malawi cichlids. Almost all the species of this complex are characterised by the single shared characteristic that they are strongly bound to the substrate offered by the rocky coast. They do, however, also have morphological features in common, that have arisen from their shared lifestyle. Thus all species have numerous small, deeply-embedded, scales on the breast region, an adaption to the sharped-edged substrate, contact with which could easily result in the loss of more loosely attached scales. This scalation changes abruptly to larger scales on the flanks. Most mbuna species (not all mbuna, some individuals do not, including almost all female *Melanochromis auratus*) have true eggspots on the anal fin, ocellated in males. An anatomical peculiarity is seen in the atrophication of the left ovary.

Depending on one's view of what constitutes a species, up to 1,600 species of cichlids are native to Lake Malawi. It should thus be obvious that it is impossible for a single scientist to undertake a complete revision of all these species. In consequence each paper dealing with a group of these fishes forms part of a patchwork, and will almost inevitably contain errors. We have to live with this problem, as only thus will there be a basis for discussion leading to further work.

The Genera (in chronological order)

Gephyrochromis Boulenger, 1901, was the first genus to be erected specifically for mbuna. Its special characters are the commonly conical form of the teeth in the outer row in both jaws, as well as the fact that the individual teeth are longer and more slender than in the closely-related genus *Pseudotropheus.* That apart, there are many similarities between the two genera. In the aquarium hobby little is known of Gephyrochromis although occasional importations of these fishes are not completely unknown. In the lake they are found singly and also in larger groups.

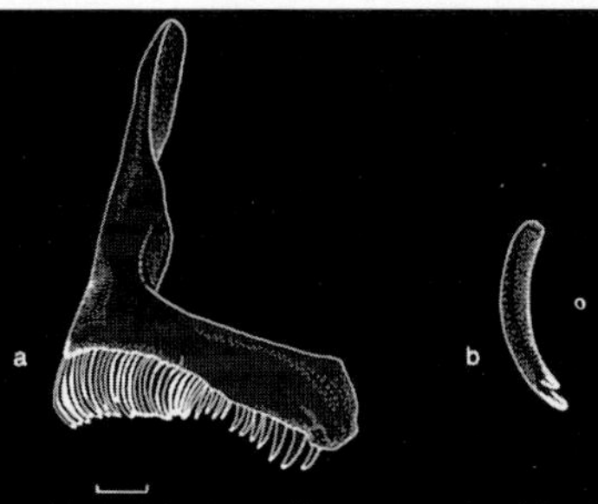

(a) Seitliche Ansicht des linken Premaxillare von *Genyochromis lawsi*; (b) Ein Zahn aus der vorderen äußeren Zahnreihe von *G. lawsi*. (Maßstrich: 1 mm)
(a) Lateral aspect of left premaxilla of Gephyrochromis lawsi; *(b) Anterior outer row tooth of* G. lawsi. Scale: 1 mm
from: Ribbink et. al., 1983

Cynotilapia and *Pseudotropheus* were both described by Regan, 1922. In both cases the type species had already been described by Günther in 1893 as members of the then catch-all cichlid genera *Hemichromis* and *Chromis,* respectively.

(a) *Cynotilapia afra*, Likoma Island, 93 mm SL; (b) Seitliche Ansicht des linken Premaxillare von *Cynotilapia afra*; (c) Ein Zahn aus der vorderen äußeren Zahnreihe von *Cynotilapia afra*.
(a) Cynotilapia afra, *Likoma Island, 93 mm SL; (b) Lateral aspect of left premaxilla of* Cynotilapia afra; *(c) Anterior outer row tooth of* Cynotilapia afra.
from: Ribbink et. al., 1983

territorial. Eine Vergesellschaftung mit anderen Mbunas in einem gut besetzten Aquarium verringert den Druck auf einzelne unterlegene Tiere. Weibchen und Jungtiere bilden in der Natur Schulen im offenen Wasser.

Die Gattung *Pseudotropheus* Regan, 1922 (die entsprechende Arbeit erschien zwar im 1921er Jahrgang, wurde aber erst 1922 veröffentlicht) wurde schon in der grundlegenden Arbeit von Ribbink et. al. (1983) in verschiedene Komplexe aufgeteilt. So kennt man neben dem *Pseudotropheus williamsi* - Komplex, der die Typusart der Gattung enthält und somit auch als *Pseudotropheus* s. str. (im eigentlichen Sinne) bezeichnet wird, noch die *Pseudotropheus elongatus*-Artgruppe, die *Ps.* "Aggressive"-Artgruppe und weitere Arten, die als *Ps.* "Miscellaneous"-Artgruppe bezeichnet werden. Der frühere *Pseudotropheus zebra*-Komplex erhielt kürzlich (Stauffer et. al., 1997) Gattungsrang und wird daher weiter hinten besprochen. Für den *Pseudotropheus tropheops*-Komplex, die letzte der sechs von Ribbink et. al. als gut abgrenzbar erkannte Artengruppe, schlug Trewavas bereits 1984 die Untergattung *Tropheops* vor. Meines Erachtens ist es nur eine Frage der Zeit, bis dieser Komplex Gattungsrang erhält. Auch auf diese Arten wird weiter hinten noch ausführlicher eingegangen. Insgesamt müssen sich die Aquarianer darauf einstellen, daß hier neue Gattungsnamen in absehbarer Zukunft auf sie zukommen.

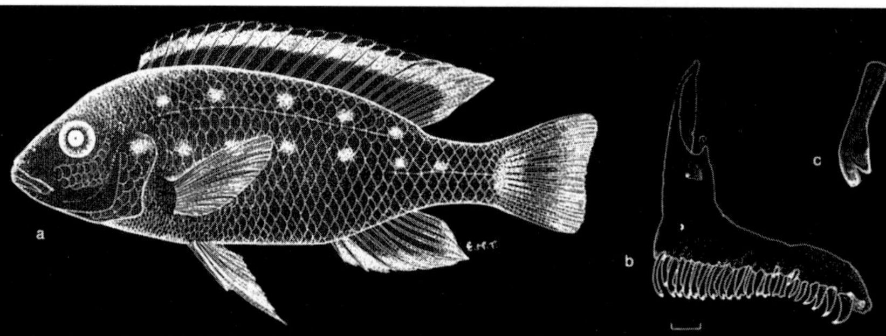

(a) *Pseudotropheus williamsi* "nkudzi", Nkudzi Point, 121 mm SL; (b) Seitliche Ansicht des linken Premaxillare von *Pseudotropheus williamsi* "nkudzi"; (c) Ein Zahn aus der vorderen äußeren Zahnreihe von *Pseudotropheus williamsi* "nkudzi". (Maßstrich: 1 mm)
(a) Pseudotropheus williamsi *"nkudzi", Nkudzi Point, 121 mm SL; (b) Lateral aspect of left premaxilla of* Pseudotropheus williamsi *"nkudzi"; (c) Anterior outer row tooth of* Pseudotropheus williamsi *"nkudzi"*. Scale: 1 mm
from: Ribbink et. al., 1983

Pseudotropheus s. str. läßt sich den anderen Komplexen nicht gut aufgrund der Bezahnung unterscheiden, vielmehr ist die Lage des Unterkiefers bei geschlossenem Maul maßgebend für die morphologische Differenzierbarkeit (bei *Pseudotropheus* s. str. verläuft der Unterkiefer parallel zu einer gedachten Linie, deren Endpunkte die Schnauzenspitze und die Hypuralplatte - bestimmte Knochen am Schwanzstiel - bilden). Aquaristisch spielt der Artkomplex

Representatives of the genus *Cynotilapia* have long, well-separated, unicuspid teeth, strongly conical in the outer row. The teeth of the inner rows are smaller and irregularly spaced. In the aquarium, as in the wild, males are highly territorial. Housing them with other mbuna in a well-populated aquarium will minimise the effect on individual subordinate fishes. Females and juveniles form schools in open water in the wild.

The genus *Pseudotropheus* Regan 1922 (the paper in question first appeared in 1921 but was first published in 1922) was first divided up into various species complexes in the ground-breaking work of Ribbink et al (1983). Thus as well as the *Ps. williamsi* complex, which contains the type species of the genus and is therefore regarded herein as *Pseudotropheus* sensu stricto (in the exact sense), it also contains the *Ps. elongatus* species group, the *Ps.* "Aggressive" species group, and other species designated as the *Ps.* "Miscellaneous" species group. Recently the group formerly known as the *Ps. zebra* complex has been raised to generic rank (Stauffer et al, 1997), more of which below. As long ago as 1984 Trewavas proposed the sub-genus *Tropheops* for the *Ps. tropheops* complex, the last of the six species groups which Ribbink et al regarded as well-defined. In my opinion it is only a matter of time before *Tropheops* is raised to generic status. These species will also be discussed further below. Generally speaking aquarists must accept the fact that there will be new generic names to contend with in the foreseeable future.

Pseudotropheus sensu stricto cannot be clearly differentiated from other genera on the basis of dentition; instead the position of the lower jaw (when the mouth is shut) is more important as regards morphological differentiation (in *Pseudotropheus* the lower jaw runs parallel to an imaginary line whose end points are the tip of the snout and

keine große Rolle auch wenn gelegentlich Vertreter eingeführt werden. Die Fische gelten als weniger aggressiv als andere Mbunas obwohl die Männchen sich untereinander heftig um Reviere streiten.

Für die drei übrigen Artengruppen, also den *Ps. elongatus*-, den *Ps.* "Aggressive"- und den *Ps.* "Miscellaneous"-Komplex, kann keine einheitliche Pflegeanleitung gegeben werden. Speziell der *Ps.* "Miscellaneous"-Komplex ist ja eine uneinheitliche Sammelgruppe. Wir verweisen daher an dieser Stelle auf die entsprechenden Symbole unter den Abbildungen im Bildteil dieses Buches. Für die Angehörigen des *Ps. elongatus*-Komplexes soll aber erwähnt werden, daß viele (aber eben nicht alle) dieser Arten trotz ihrer relativ geringen Körpergröße recht rabiat untereinander werden können und daher verhältnismäßig große Aquarien brauchen. In noch viel stärkerem Maße gilt das für die *Ps.* "Aggressive"- Gruppe. Hier muß man beachten: nomen est omen. Die Gruppe erhielt von Ribbink et.al. ihre Bezeichnung, weil die Fische im See ihr Revier derart vehement gegen andere Fische verteidigen, daß sogar regelrechte Algenrasen zu wachsen vermögen!

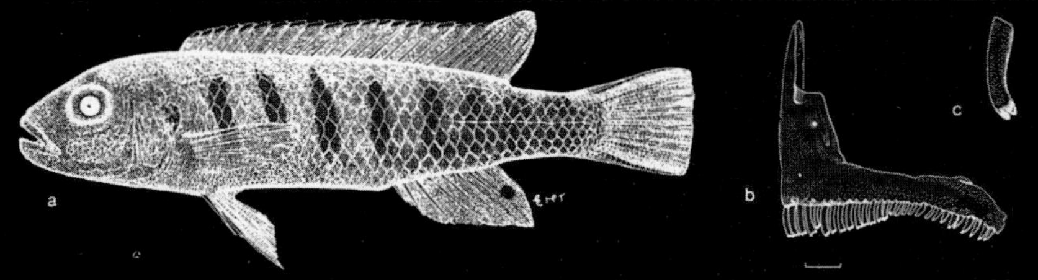

(a) *Pseudotropheus elongatus*, Nkhata Bay, 80 mm SL; (b) Seitliche Ansicht des linken Premaxillare von *Pseudotropheus elongatus* "black" von Chinyankwazi I.; (c) Ein Zahn aus der vorderen äußeren Zahnreihe von *P. elongatus* "black" von Chinyankwazi I.. (Maßstrich: 1 mm)
(a) Pseudotropheus elongatus, *Nkhata Bay, 80 mm SL; (b) Lateral aspect of left premaxilla of* Pseudotropheus elongatus *"black" from Chinyankwazi I.; (c) Anterior outer row tooth of* Pseudotropheus elongatus *"black" from Chinyankwazi I.*. Scale: 1 mm
from: Ribbink et. al., 1983

Labeotropheus Ahl, 1926 ist eine relativ einfach zu erkennende Gattung, weil ihre Arten durch den Besitz einer fleischigen Verlängerung der Schnauze gekennzeichnet sind. Ihr Maul ist beinahe unterständig und dadurch sind die Fische in der Lage den Algenrasen abzugrasen, während sie parallel zur Felsoberfläche schwimmen. *Labeotropheus* ist aquaristisch gut bekannt. In nicht zu kleinen Behältern lassen sich die Arten gut mit anderen Mbunas vergesellschaften. In größeren Aquarien sind auch mehrere Männchenreviere möglich.

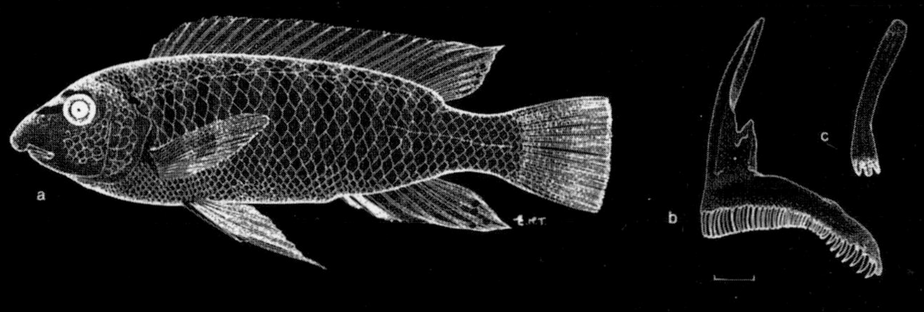

(a) *Labeotropheus trewavasae*, Maleri Island, 90 mm SL; (b) Seitliche Ansicht des linken Premaxillare von *Labeotropheus trewavasae*; (c) Ein Zahn aus der vorderen äußeren Zahnreihe von *Labeotropheus trewavasae*. (Maßstrich: 1 mm)
(a) Labeotropheus trewavasae, *Maleri Island, 90 mm SL; (b) Lateral aspect of left premaxilla of* Labeotropheus trewavasae; *(c) Anterior outer row tooth of* Labeotropheus trewavasae. Scale: 1 mm
from: Ribbink et. al., 1983

Trewavas hat 1935 in einer Pionierarbeit über die Cichliden des Malawisees gleich fünf neue Mbuna-Gattungen aufgestellt.

the hypural plate (a group of bones in the caudal peduncle). Aquaristically speaking this species complex is relatively unimportant, even though representatives of it are occasionally imported. They are regarded as less aggressive than other mbuna, although males are highly territorial among themselves. No all-embracing hints on maintenance can be given for the three remaining species groups, ie the *Ps. elongatus*, the *Ps.* "Aggressive", and the *Ps.* "Miscellaneous" complexes. In particular the *Ps.* "Miscellaneous" complex is a diverse assemblage. We thus instead advise reference to the relevant symbols beneath the photos of the individual species in the pictorial part of this book. As regards the members of the *Ps. elongatus* complex, however, it should be noted that many (but not all) can be very quarrelsome among themselves, despite their relatively small size, and therefore require relatively large aquaria. This applies in even greater measure to the *Ps.* "Aggressive" group, where the saying "nomen is omen" is truly applicable. Ribbink et al gave this group its name because in the lake these cichlids defend their territories so violently against other fishes that algae can grow into regular "gardens" there!

Labeotropheus Ahl, 1926 is a relatively easily recognisable genus, as its species are characterised by a fleshy prolongation of the snout.

The mouth is underslung, and hence these fishes are able to graze algae while swimming parallel to the rock surface. *Labeotropheus* is well known to aquarists. The species can readily be housed with other mbuna in tanks of reasonable size. Several male territories are possible in larger aquaria.

In 1935 Trewavas erected 5 new mbuna genera simultaneously in her pioneering work on the cichlids of Lake Malawi.

Cyathochromis Trewavas, 1935 has remained monotypic to the present day. Externally *C. obliquidens*

Cyathochromis TREWAVAS, 1935 ist bis zum heutigen Tag monotypisch geblieben. *C. obliquidens* ähnelt äußerlich einer Art des *Maylandia zebra*-Komplexes, besitzt aber eine völlig abweichende Bezahnung mit schlanken Zahnhälsen und etwas zusammengedrückt wirkenden, löffelförmigen Kronen, die in der äußeren Reihe gleichförmig schräg vom Knochen wachsen. Im Aquarium wird die Art nur sehr selten gepflegt, Zuchtberichte aus der Gefangenschaft

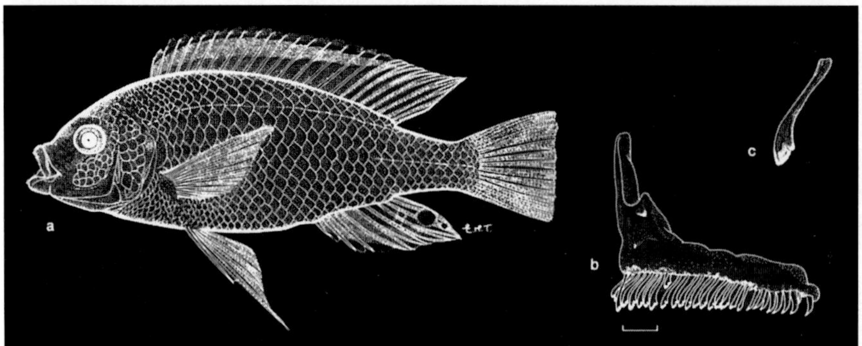

(a) *Cyathochromis obliquidens*, Monkey Bay, 102 mm SL; (b) Seitliche Ansicht des linken Premaxillare von *Cyathochromis obliquidens*; (c) Ein Zahn aus der vorderen äußeren Zahnreihe von *Cyathochromis obliquidens*. (Maßstrich: 1 mm)
(a) Cyathochromis obliquidens, *Monkey Bay, 102 mm SL; (b) Lateral aspect of left premaxilla of* Cyathochromis obliquidens; *(c) Anterior outer row tooth of* Cyathochromis obliquidens. Scale: 1 mm

from: Ribbink et. al., 1983

fehlen. Im See werden vom Männchen Laichgruben zwischen Pflanzen oder unter Steinen gegraben.

(a) *Genyochromis mento*, Monkey Bay, 85 mm SL; (b) Seitliche Ansicht des linken Premaxillare von *Genyochromis mento*; (c) Ein Zahn aus der vorderen äußeren Zahnreihe von *Genyochromis mento*. (Maßstrich: 1 mm)
(a) Genyochromis mento, *Monkey Bay, 85 mm SL; (b) Lateral aspect of left premaxilla of* Genyochromis mento; *(c) Anterior outer row tooth of* Genyochromis mento. Scale: 1 mm

from: Ribbink et. al., 1983

Genyochromis TREWAVAS, 1935 ist eine weitere monotypische Gattung, die allerdings in einer großen Variabilität bekannt ist. Deshalb wurde schon vermutet, daß sich mehrere Arten hinter dem Taxon *G. mento* verbergen. Sie ist recht gut an dem vorstehenden Unterkiefer und dem breiten Maul zu erkennen. Sie ist der einzige Mbuna aus dem Malawisee, die nur schlecht mit anderen Fischen vergesellschaftet werden kann, weil sie eine sehr spezialisierte Ernährungsweise besitzt, die sie auch im Aquarium nicht ändert: Sie ist Schuppen- und Flossenfresser. Angeblich werden nur die Angehörigen der eigenen Art nicht angefallen. Es scheint möglich zu sein, diese Tiere im Artbecken zu pflegen.

Labidochromis TREWAVAS, 1935 ist eine Gattung, die durch lange, spitze, einspitzige, nach vorne gerichtete vordere Zähne gekennzeichnet ist. Dadurch unterscheidet sie sich deutlich von ähnlichen Formen, die zur Zeit noch der Gattung *Pseudotropheus* sensu lato zugerechnet werden. Die Typusart ist *L. vellicans* TREWAVAS, 1935. Besonders LEWIS hat die Gattung bei seiner Revision 1982 um zahlreiche Arten erweitert, von denen *L. mylodon* umstritten ist, da sie sich nur durch die mühlsteinartige Pharyngealbezahnung morphologisch von *L. pallidus* unterscheidet. Wissenschaftliche Untersuchungen bei anderen Haplochrominen haben aber ergeben, daß die Ausprägung der Schlundzähne in einer sensiblen Phase während der Ontogenese (der Individual-

looks like a member of the *Maylandia zebra* complex, but possesses a totally different dentition, the teeth being slender at their bases and giving a somewhat close-packed impression, with spoon-shaped crowns; those of the outer row are set in the bone at a uniform angle. This species is only very rarely maintained in the aquarium, and there are no reports of captive breeding. In the lake males dig spawning pits between plants or beneath stones.

Genyochromis TREWAVAS, 1935 is another monotypic genus, which is, however, noted for its great variability. For this reason it has been suggested that the taxon *G.mento* actually includes several species. The species is easily recognised by its protrusive lower jaw and broad mouth. It is the only mbuna species from Lake Malawi that is virtually impossible to keep with other fishes, as it has a highly specialised feeding method, which it retains in the aquarium: it is a scale and fin eater. It is said that only conspecifics are safe from attack; hence it may be possible to keep these fishes in a species aquarium.

Labidochromis TREWAVAS, 1935 is characterised by long, pointed, unicuspid, forward-pointing teeth, and is thus clearly distinct from similar forms which are at present still assigned to *Pseudotropheus* sensu lato. The type species is *L. vellicans* TREWAVAS, 1935. The genus has since been enlarged, especially by LEWIS in his 1982 revision, to include numerous species, of which *L. mylodon* is questionably valid, as its only morphological difference from *L. pallidus* is its millstone-like pharyngeal dentition. Scientific investigation of other haplochromines has shown, however, that the development of the pharyngeal teeth is affected by the diet during a critical phase of ontogeny, and is thus not to be regarded as species-specific. Whether this applies to *L. mylodon / L. pallidus* has not been researched. Many *Labidochromis* are only slightly or not at all territorial. In nature some species (eg *L. heterodon*) are found in large shoals. Others, such as full-grown males of *L. caeru-*

(a) *Labidochromis shiranus*, Nkudzi Point, 66 mm SL; (b) Seitliche Ansicht des linken Premaxillare von *Labidochromis vellicans*; (c) Ein Zahn aus der vorderen äußeren Zahnreihe von *Labidochromis vellicans*. (Maßstrich: 1 mm)
(a) Labidochromis shiranus, *Nkudzi Point, 66 mm SL; (b) Lateral aspect of left premaxilla of* Labidochromis vellicans; *(c) Anterior outer row tooth of* Labidochromis vellicans. Scale: 1 mm

from: Ribbink et. al., 1983

entwicklung) durch die Art der Nahrung festgelegt wird, also nicht alleine arttypisch sein muß. Ob dies auch für *L. mylodon/L. pallidus* gilt, wurde noch nicht untersucht. Viele *Labidochromis* sind wenig oder nicht territorial. Manche Arten treten in der Natur in großen Schulen auf (z.B. *L. heterodon*). Andere, wie *L. caeruleus* sind im Aquarium sehr wohl territorial, obwohl aus Seebeobachtungen anderes berichtet wird. Besonders ältere Männchen können nur in größeren Aquarien und nur unter ständigen Streitigkeiten zusammen gepflegt werden, wobei unterlegene Tiere immer wieder lädiert werden.

Melanochromis TREWAVAS, 1935 ist eine weitere aquaristisch gut bekannte Gattung, die ursprünglich anhand von Unterschieden in den Schlundzähnen von *Pseudotropheus* abgegrenzt wurde. Erst viel später wurde vor allem in populärwissenschaftlichen Veröffentlichungen das Farbkleid, insbesondere die längsgestreifte Grundstruktur der Zeichnung, zu dem entscheidenden Kriterium, Fische zu *Melanochromis* oder *Pseudotropheus* zu stellen. Dabei wurden manche Arten von der einen in die andere Gattung transferiert. LEWIS fand später weitere Argumente (etwa die Umkehrung des hell/dunkel Bandes bei männlichen/weiblichenTieren) um eine Trennung nach farblichen Kriterien auch wissenschaftlich zu untermauern. Leider kam es bei der Beschreibung von einigen *Melanochromis*-Arten zu Ungereimtheiten, wie etwa falsch vermittelten Fundortangaben. Dies bewirkte, daß einige Arten in der Folgezeit verwechselt, doppelt beschrieben oder wissenschaftlich ignoriert wurden. Bei der Beschreibung von *M. heterochromis* BOWERS & STAUFFER, 1993 wurde anscheinend übersehen, daß sich die Proportionen bei in Konservierungsflüssigkeit befindlichen Exemplaren im Laufe der Jahrzehnte ändern können. Der Name muß deshalb als Synonym von *M. vermivorus* angesehen werden,

leus, are very territorial in the aquarium although observations in the lake indicate otherwise. In particular older males can be kept only in larger aquaria and will constantly battle with one another, such that low-ranking individuals are repeatedly injured.

Melanochromis TREWAVAS, 1935 is another genus well-known to aquarists, which was originally differentiated from *Pseudotropheus* on the basis of differences in the pharyngeal dentition. Much later, in particular in the popular scientific publications, coloration, especially the basic pattern of longitudinal bands, was used as the criterion for assigning fishes to *Melanochromis* or *Pseudotropheus*, with some species being transferred from one genus to the other. Later LEWIS discovered additional arguments (for example the reversed dark / light banding in males / females which scientifically led to differentiation according to coloration-based criteria. Unfortunately a number of absurd errors were made in the descriptions of a few *Melanochromis* species, for example in-correct locality data. As a result some species were subsequently confused, described a second time, or scientifically ignored. The description of *M. heterochromis* BOWERS & STAUFFER, 1993 apparently failed to take into account the fact that over the course of decades the proportions of preserved specimens can alter. The name must therefore be regarded as a synonym of *M. vermivorus*, likewise *M. mellitus* JOHNSON, 1976. *M. loriae* JOHNSON, 1975 was commonly identified in the aquarium literature as *M. chipokae*, although the original description included colour photos unmistakeably identifying the species. *M. chipokae* JOHNSON, 1975 is, like *M. robustus* JOHNSON, 1976, a rather dubious species. Whether or not the latter is identical with *M. benetos* BOWERS & STAUFFER, 1997, is a moot point. *M. elastodema* BOWERS & STAUFFER, 1997 was not compared with *M. interruptus* JOHNSON, 1975

(a) *Melanochromis melanopterus*, Monkey Bay, 126 mm SL; (b) Seitliche Ansicht des linken Premaxillare von *Melanochromis melanopterus*; (c) Ein Zahn aus der vorderen äußeren Zahnreihe von *Melanochromis melanopterus*. (Maßstrich: 1 mm)
(a) Melanochromis melanopterus, *Monkey Bay, 126 mm SL; (b) Lateral aspect of left premaxilla of* Melanochromis melanopterus; *(c) Anterior outer row tooth of* Melanochromis melanopterus. Scale: 1 mm

from: Ribbink et. al., 1983

ebenso gesellt sich *M. mellitus* JOHNSON,1976 hierher. *M. loriae* JOHNSON, 1975 wurde in der aquaristischen Literatur häufig als *M. chipokae* bezeichnet, obwohl es zur Erstbeschreibung Farbfotos gab, welche die Art unverwechselbar erkennen lassen. *M. chipokae* JOHNSON, 1975 ist ebenso wie *M. robustus* JOHNSON, 1986 eine etwas dubiose Art. Ob letztere mit *M. benetos* BOWERS & STAUFFER, 1997 identisch ist, ist fraglich. *M. elastodema* BOWERS & STAUFFER, 1997 wurde in der Erstbeschreibung nicht einmal mit *M. interruptus* JOHNSON, 1975 verglichen, obwohl es die einzige andere Art mit demselben Farbkleid war. Heute müssen sie als Synonym betrachtet werden. Die Gattung *Melanochromis* wird von modernen Ichthyologen als polyphyletisch angesehen, daß heißt, manche Arten werden in absehbarer Zeit noch von ihr abgespalten werden. Im Aquarium sind viele Arten (z.B. *M. auratus*) sehr aggressiv. Nur in größeren Behältern können mehrere Männchen zusammen gepflegt werden, oder die Tiere müssen in einem deutlich überbesetzten Aquarium gepflegt werden, damit sich die Aggressionen verteilen.

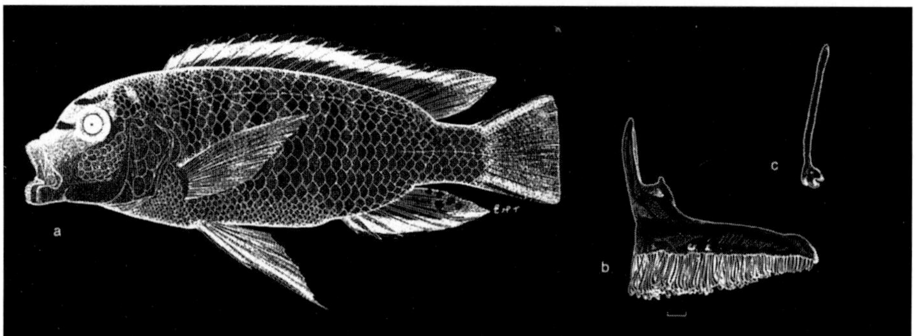

(a) *Petrotilapia* "mumbo blue", Nkhata Bay, 80 mm SL; (b) Seitliche Ansicht des linken Premaxillare von *Pseudotropheus elongatus* "black" von Chinyankwazi I.; (c) Ein Zahn aus der vorderen äußeren Zahnreihe von *P. elongatus* "black" von Chinyankwazi I.. (Maßstrich: 1 mm)
(a) Pseudotropheus elongatus, *Nkhata Bay, 80 mm SL; (b) Lateral aspect of left premaxilla of* Pseudotropheus elongatus *"black" from Chinyankwazi I.; (c) Anterior outer row tooth of* Pseudotropheus elongatus *"black" from Chinyankwazi I..* Scale: 1 mm
from: Ribbink et. al., 1983

Petrotilapia TREWAVAS, 1935 ist die Mbuna-Gattung mit den größten Arten. Das Maul dieser Fische ist endständig und bei erwachsenen Tieren stets leicht geöffnet. Besonders auffällig sind die gut sichtbaren, zahlreichen, langen, schlanken, dreispitzigen Zähne die in jedem Kiefer stehen und den Fischen ein Aussehen verleihen, als hätten sie geschwollene Lippen. *Petrotilapia* ist aufgrund der Bezahnung besonders gut dafür ausgerüstet, den Aufwuch von den Felsen abzuweiden (sie tun dies, indem sie ihr Maul im rechten Winkel zur Felsoberfläche pressen). Dennoch wurde schon beobachtet, daß sie bei einem reichen Planktonangebot auch dieses aus dem freien Wasser picken. *Maylandia fainzilberi* hat als einziger Nicht-*Petrotilapia* eine sehr ähnliche Bezahnung und wird von manchen Autoren deshalb mit einem Sonderstatus bedacht. Da die Art auf Gattungsebene ansonsten aber keine Unterschiede zu den übrigen *Maylandia*-Arten besitzt, wird hier die Auffassung vertreten, sie in *Maylandia* zu integrieren. *Petrotilapia*-Männchen besetzen in der Natur sehr große Reviere (für manche Arten wurden über 20 Quadratmeter gemessen), weshalb sie im Aquarium als problematisch gelten. Weibchen und Jungtiere hingegen schließen sich zu großen Schulen zusammen.

Iodotropheus OLIVER & LOISELLE, 1972 ist eine kleine Gattung, die einen leicht vorstehenden Unterkiefer besitzt und ein Häutchen hat, das die Oberlippe mit der Schnauze verbindet. Die Zähne der äußeren Reihe sind zweispitzig, die der Inneren dreispitzig. *I. declivitas* STAUFFER, 1994 wird von KONINGS als

at the time of its original description, although the latter is the only other species with a similar colour pattern. They must now be regarded as synonymous. The genus *Melanochromis* is regarded as polyphyletic by modern ichthyologists, ie at some time in the foreseeable future some species will be split off from it. In the aquarium many species (eg *M. auratus*) are very aggressive; several males can be kept together only in larger tanks, or else they must be kept in a significantly over-populated aquarium in order to spread the aggression.

Petrotilapia TREWAVAS, 1935 is the mbuna genus with the largest species. The mouth of these fishes is terminal, and in adult individuals always slightly open. The numerous clearly visible, slender, tricuspid teeth in both jaws are particularly striking, and give the fishes the appearance of having swollen lips. On account of its dentition *Petrotilapia* is particularly well equipped for grazing algae from rocks (they do this with their mouths at right-angles to the rock surface). However, it has been observed that when plankton is in plentiful supply they also pick this from the water column. *Maylandia fainzilberi* is the only non-*Petrotilapia* with a very similar dentition, and is thus afforded special status by some authors, but as it otherwise exhibits no differences (on generic level) from other *Maylandia* species, the view is here taken that it should be included in *Maylandia*. In nature *Petrotilapia* males occupy very large territories (in some species measured at more than 20 square metres), and for this reason they can be problematical in the aquarium. Females and juveniles, on the other hand, form large shoals.

Iodotropheus OLIVER & LOISELLE, 1972 is a small genus with a slightly protruding lower jaw and a frenum, that is, a small membrane joining the upper lip and the snout centrally. The teeth of the outer rows are bicuspid, those of the inner tricuspid. *I. declivitas* STAUFFER, 1994 is regarded by KONINGS (1995) as a synonym of *I. sprengerae*, although there are differences in the body proportions of the two populations. The key to the understanding of such differences of opinion is the different view of what constitutes a species held by each author. See the contribution by KELLOG and STAUFFER, "A species definition for Malawi Cichlids", in this volume. *Iodotropheus* are regarded as relatively unaggressive and can thus easily be kept together in the aquarium.

Synonym zu *I. sprengerae* betrachtet, obwohl es Unterschiede in den Körperproportionen beider Populationen gibt. Der Schlüssel zum Verständnis dieser Meinungsverschiedenheiten ist das unterschiedliche Artkonzept, das die jeweiligen Autoren vertreten. Lesen Sie hierzu den Beitrag von KELLOG & STAUFFER "Eine Artdefinition für die Malawisee-Cichliden" in diesem Band. *Iodotropheus* gelten als wenig aggressiv und sind deshalb im Aquarium gut zu vergesellschaften.

(a) *Iodotropheus sprengerae*, Chinjankwazi Island, 82 mm SL; (b) Seitliche Ansicht des linken Premaxillare von *Iodotropheus sprengerae;* (c) Ein Zahn aus der vorderen äußeren Zahnreihe von *Iodotropheus sprengerae.* (Maßstrich: 1 mm)
(a) Iodotropheus sprengerae,*Chinyankwazi Island, 82 mm SL; (b) Lateral aspect of left premaxilla of* Iodotropheus sprengerae*;*
(c) Anterior outer row tooth of Iodotropheus sprengerae. Scale: 1 mm
from: Ribbink et. al., 1983

Microchromis JOHNSON, 1975 ist eine monotypische Gattung, die von vielen Autoren ignoriert oder als Synonym von *Cynotilapia* betrachtet wird. Daran mag der simple Umstand schuld sein, daß der Trivialname "Zwergzebra" für die beiden Arten *M. zebroides* und *C. afra* verwendet wird und sie deshalb miteinander verwechselt wurden. Bereits JOHNSON geht in seiner Gattungsbeschreibung auf die Unterschiede zu *Cynotilapia* ein, z.B. die nur zwei Zahnreihen, unterschiedliche Kopfproportionen und andere Schuppenzahlen. Auch das der Erstbeschreibung beigefügte Farbfoto zeigt einen von *Cynotilapia* verschiedenen Fisch, weshalb die Gattung und Art *M. zebroides* hier als valide betrachtet werden.

1984 haben TREWAVAS einerseits und MEYER & FOERSTER andererseits in zwei verschiedenen Abhandlungen, die in der selben Zeitschrift erschienen sind, Untergattungen zu *Pseudotropheus* beschrieben. Durch die Beschreibung von *Metriaclima* STAUFFER & al., 1997 als Gattungsnamen für Arten des Pseudotropheus zebra-Komplexes ist die von MEYER & FOERSTER beschriebene Untergattung MAYLANDIA automatisch in den Gattungsrang erhoben worden, weil ihre Typusart (*Pseudotropheus greshakei*), ein Mitglied des *P. zebra*-Komplexes ist. *Metriaclima* ist nun nur ein jüngeres Synonym zu *Maylandia*. STAUFFER & al., haben zwar in der Beschreibung von *Metriaclima* moniert, daß *Maylandia* ein nomen nudum sei, diese Behauptung konnte

Microchromis JOHNSON, 1975 is a monotypic genus which is ignored or regarded as a synonym of *Cynotilapia* by many authors. This may be the result of the simple circumstance that the trivial name "Dwarf Zebra" has been used for both *M. zebroides* and *C. afra*, and as a result the two have been confused. However, JOHNSON indicated differences from *Cynotilapia* in his original description, eg only two rows of teeth, different head proportions, and different scale counts. Moreover the photo accompanying the original description shows a fish quite different to a *Cynotilapia,* for which reason the genus and the species *M. zebroides* are here regarded as valid.

In 1984 TREWAVAS on the one hand, and MEYER & FOERSTER on the other, described subgenera of *Pseudotropheus* in two different papers which appeared in the same journal. As a result of the description of *Metriaclima* STAUFFER et al, 1997 as a genus name for the species of the *Pseudotropheus zebra* complex, MEYER & FOERSTER's subgenus is automatically raised to generic status, as their type species (*Ps. greshakei*) is a member of that complex. *Metriaclima* is now a junior synonym of *Maylandia.* STAUFFER et al have in fact indicated, in their description of *Metriaclima,* that *Maylandia* is a nomen nudum, but this assumption is not supported by examination of the description of *Maylandia,* in which MEYER & FOERSTER refer to explicit differences in the dentition and melanin patterns of their subgenus

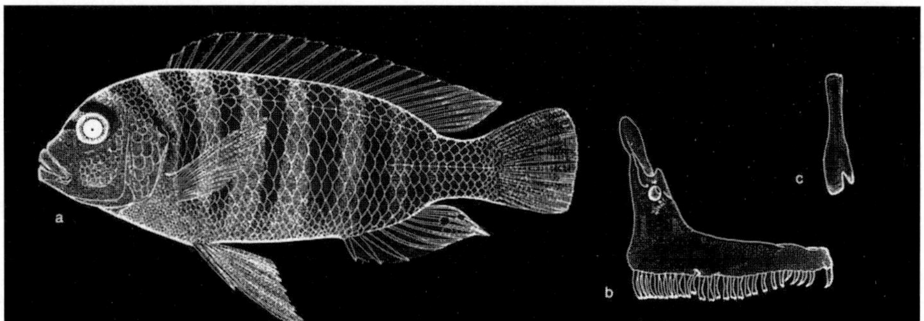

(a) *Maylandia zebra*, Monkey Bay, 95 mm SL; (b) Seitliche Ansicht des linken Premaxillare von *Maylandia zebra;* (c) Ein Zahn aus der vorderen äußeren Zahnreihe von *Maylandia zebra.* (Maßstrich: 1 mm)
(a) Maylandia zebra, *Monkey Bay, 95 mm SL; (b) Lateral aspect of left premaxilla of* Maylandia zebra*; (c) Anterior outer row tooth of* Maylandia zebra. Scale: 1 mm
from: Ribbink et. al., 1983

aber einer Überprüfung der *Maylandia*-Beschreibung nicht standhalten. MEYER & FOERSTER haben darin u. a. explizit Unterschiede in der Bezahnung und der Melaninfärbung ihrer Untergattung zur Typusart von *Pseudotropheus* (*P. williamsi*) hervorgehoben. Deshalb wird in diesem Band der Gattungsname *Maylandia* für Vertreter des P. zebra-Komplexes verwendet. Nach STAUFFER & al., sind die Angehörigen dieses Komplexes (innerhalb *Pseudotropheus* s.l.) durch eine mäßig steile Stirn, einer Verdickung im Knochen des nasenartigen Schädelfortsatzes und einem etwa 45° winkligen Unterkiefer zu einer gedachten Linie zwischen Schnauzenspitze und Hypural (besonderen Knochen im Schwanzstiel) gekennzeichnet. *Maylandia*-Arten sind in der Aquaristik weit verbreitet und sehr beliebt. Die Arten sind mäßig aggressiv, lassen sich in nicht zu kleinen und entsprechend eingerichteten Aquarien aber problemlos vergesellschaften.

Analog zu *Maylandia* wird die von TREWAVAS 1984 als Untergattung eingeführte Bezeichnung *Tropheops* für Vertreter des *Pseudotropheus tropheops*-Komplexes wahrscheinlich in Kürze in den Gattungsrang erhoben. Obwohl der Winkel mit dem das Stirn-Profil ansteigt von Art zu Art etwas unterschiedlich ist, ist er bei allen Angehörigen dieser (noch) Untergattung besonders steil (am flachsten bei *Ps.* (*T.*) cf. *novemfasciatus*). Der Schädelknochen hat einen relativ kurzen rostralen Fortsatz. Von Arten des *Pseudotropheus elongatus*-Komplexes mit einem ähnlichen Kopfprofil, lassen sich erwachsene Tiere durch eine wesentlich größere Körperhöhe im Vergleich zur Körperlänge unterscheiden. Einer ausführlicheren Abgrenzung sollte hier nicht vorgegriffen werden. *Tropheops*-Männchen sind territorial und nur in größeren Aquarien lassen sich mehrere Männchen zusammen pflegen. Im See wurden teils tageszeitlich abhängige Revierverteidigungszeiten festgestellt, außerhalb derer auch Männchen friedlich nebeneinander bei der Nahrungssuche beobachtet wurden.

Zusammen mit der Gattungsbeschreibung von *Metriaclima*, haben STAUFFER, BOWERS, KELLOG & MCKAYE zehn neue Arten des früheren *Pseudotropheus zebra*-Komplexes beschrieben, die aufgrund der vorstehenden Erläuterungen in diesem Band alle in der Gattung *Maylandia* eingeordnet wurden. Die Auffassung der Autoren, *Pseudotropheus fainzilberi* wegen seiner an *Petrotilapia* erinnernden Bezahnung einen Sonderstatus zu geben, wird in diesem Band nicht geteilt und die Art zu *Maylandia* gestellt. Die oben genannten Autoren haben auch *Pseudotropheus heteropictus* STAECK in ihre Gattung *Metria*-

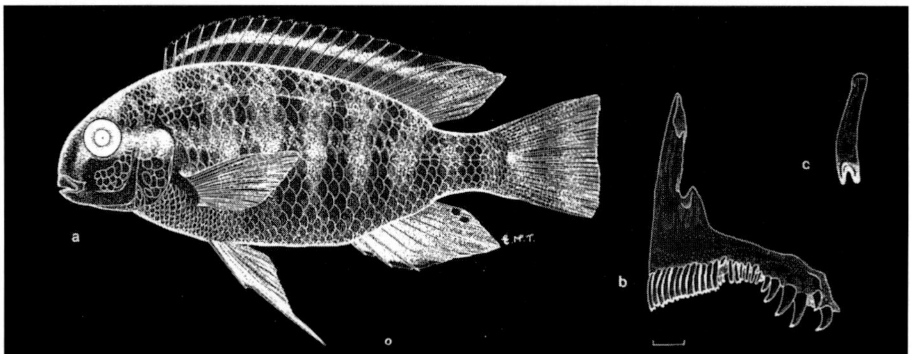

(a) *Pseudotropheus (Tropheops) tropheops* "Orange Chest", Monkey Bay, 88 mm SL; (b) Seitliche Ansicht des linken Premaxillare von *Pseudotropheus (Tropheops) tropheops* "Orange Chest"; (c) Ein Zahn aus der vorderen äußeren Zahnreihe von *Pseudotropheus (Tropheops) tropheops* "Orange Chest". (Maßstrich: 1 mm)

(a) Pseudotropheus (Tropheops) tropheops *"orange chest", Monkey Bay, 88 mm SL; (b) Lateral aspect of left premaxilla of* Pseudotropheus (Tropheops) tropheops *"orange chest"; (c) Anterior outer row tooth of* Pseudotropheus (Tropheops) tropheops *"orange chest"*. Scale: 1 mm

from: Ribbink et. al., 1983

compared to those of the type species of *Pseudotropheus* (*Ps. williamsi*). For this reason we have used the generic name *Maylandia* for members of the *Ps. zebra* complex in this volume. According to STAUFFER et al the members of this complex (in opposite to other members of *Pseudotropheus* s.l.) are characterised by their skull morphology (specifically a moderately-sloped ethmo-vomerine block and a swollen rostral tip) and a lower jaw that forms an angle of about 45° with a line from the tip of the snout to the hypural plate (a group of bones in the caudal peduncle). *Maylandia* species are widespread in the hobby and very popular. They are moderately aggressive, but can be kept without problem in reasonably large and properly decorated aquaria.

Analogous with *Maylandia* is the name *Tropheops*, introduced by Trewavas 1984 as a subgeneric taxon for the representatives of the Pseudotropheus tropheops complex, and which will probably shortly be elevated to generic status. Although the angle at which the upper head profile ascends is variable, in all the members of this (at present still) subgenus it is particularly steep (flattest in *Ps. (T.)* cf. *novemfasciatus*). The rostral tip of the skull is relatively short. Adults can be distinguished from species of the *Ps. elongatus* complex by virtue of an appreciably greater body depth in relation to body length, however we will not include a more extensive differentiation here. *Tropheops* males are territorial and only in larger aquaria can several males be kept together. In the lake it has been established that there are territorial periods during the daytime, but at other times males were seen peacefully searching for food together.

Together with the generic description of *Metriaclima*, STAUFFER, BOWERS, KELLOG, and MCKAYE also described 10 new species of the former *Ps. zebra* complex, which, as explained earlier, are herein assigned to the genus *Maylandia*. The view of the authors that *Ps. fainzilberi* should be given special status on account of its *Petrotilapia*-like dentition is not adopted in this book, and the species is likewise placed in *Maylandia*. The above-mentioned authors also included *Ps. heteropictus* STAECK 1980 in their genus *Metriaclima*. But although this species does have characteristics that justify such an inclusion, other characters counter the argument. Because the live appearance

clima integriert. Obwohl die Art typische Merkmale besitzt, die diese Eingruppierung rechtfertigen würde, sprechen andere Merkmale dagegen. Da die Art aufgrund ihrer Erscheinung als lebende Tiere nicht gut in *Maylandia* paßt, wird sie in diesem Band bis zur weiteren Klärung in *Pseudotropheus* belassen. Die Beschreibung der zehn neuen Arten des früheren *Pseudotropheus zebra*-Komplexes nach phylogenetischen Gesichtspunkten ist auch auf herbe Kritik gestoßen. Wegen der Aktualität der Beschreibungen, die in der aquaristischen Literatur bisher kaum Eingang gefunden haben, soll eine kritische Anmerkung von Ad Konings und eine Erwiederung durch zwei der Autoren der Original-Beschreibungen (Karen A. Kellog und Jay R. Stauffer) die dahinterstehenden Artkonzepte auch für Aquarianer transparenter machen. In beiden Aufsätzen wird *Metriaclima* und nicht *Maylandia* als Gattungsnamen verwendet, dies deckt sich aber aus schon erläuterten Gründen nicht mit der Auffassung des Autors dieses Bandes.

Die Fortpflanzung der Mbunas

Alle Mbunas sind maternale Maulbrüter, das heißt, das Weibchen ist allein für die Brutfürsorge von Eiern, Larven und Jungfischen zuständig. Es ist aber auch der Teil, der sich den geeigneten Geschlechtspartner vorher aussucht. Die Rolle der Männchen ist es, mit möglichst attraktiven Farben und Balzbewegungen auf sich aufmerksam zu machen und sie müssen genau so aussehen, wie ein Weibchen sich seinen Ablaichpartner vorstellt. Für eine optimale Fitness im Überlebenskampf ist es sinnvoll, die für eine Art wichtigsten Merkmale weiterzuvererben. Solche Merkmale sind genetisch am ehesten von Partnern der selben Art zu erwarten. Deshalb sind arterkennende Mechanismen wohl selbst zum genetisch fixierten Programm jeder Art geworden - und Abweichungen im Farbkleid vererben sich deshalb auch sofort weiter. Besonders für Mbunas ist deshalb das SMRS, (specific mate recognition system = das artspezifische Geschlechtspartner-Erkennungssystem) nach Paterson (1980), eines der wichtigsten arterhaltenden Prozesse geworden. Landläufig wird immer angenommen, daß die Färbung das ausschlaggebendste dabei ist, weil wir Menschen selbst für visuelle Reize besonders empfänglich sind. Tatsächlich ist es das Farbkleid der Männchen, das wir bei lebenden Mbunas, einige Kenntnisse vorausgesetzt, sofort zur Unterscheidung der einzelnen Arten hernehmen können. Es sind dagegen bisher kaum Untersuchungen angestellt worden, ob nicht auch bei diesen Tieren Pheromone, also Duftstoffe (Wasser ist ein hervorragender Vermittler von Geruchsstoffen und als Transportmedium wesentlich besser geeignet als Luft), die eigentlichen Reizträger sind. Wie dem auch sei, verstehen es manche Männchen laichvolle Weibchen anzulocken. Die über Felsen lebenden Mbunas sind aufgrund des Substrates, über dem sie leben, nicht sehr erfinderisch im Bau von Ablaichplätzen geworden. Im Gegenteil, das abgenagte Stück Stein, das ihr Territorium darstellt, ist gleichzeitig auch ihr Laichplatz. Das Männchen hat bei der Balz die dominierende Rolle. Es setzt sich durch sein Gehabe auch der Gefahr aus, besonders in flacheren

of the species does not accord well with *Maylandia,* it has been left in *Pseudotropheus* in this volume, pending further clarification. The description of the 10 new species of the former *Ps. zebra* complex has attracted severe criticism on phylogenetic grounds. Because of the recentness of the descriptions, which have so far hardly been mentioned in the aquarium literature, this volume includes a critical review by Ad Konings and a reply by two of the authors of the original descriptions (Karen A. Kellog and Jay R. Stauffer), in order to clarify the underlying species concepts for aquarists as well. In both items *Metriaclima* rather than *Maylandia* is used as the generic name; this does not, however, supersede the views of the authors of this volume, as detailed earlier.

Reproduction in Mbuna

All mbuna are maternal mouthbrooders, ie the female alone is responsible for the brood care of the eggs, larvae, and fry. It also falls to her, however, to seek out the correct breeding partner beforehand. The role of the male is simply to make himself as conspicuous as possible by means of attractive colours and courtship display, and at the same time to conform exactly to the female's idea of a suitable breeding partner. In order to ensure optimum fitness in the battle for survival, it makes sense for the most important characteristics to be heritable. Such genetic characters are most likely to be expected by partners belonging to the same species. For this reason species-recognition characters have become part of the genetically fixed "program" of each species - and thus aberrations in colour pattern are likewise immediately passed on. According to Paterson (1980) in mbuna in particular, therefore, the SMRS (Specific Mate Recognition System) is one of the most important processes in maintaining species integrity. It is generally assumed that coloration is the most significant factor, as humans are particularly sensitive to visual stimuli. And it is in fact male coloration which, apart from a few other characters, we initially utilise to differentiate living mbuna. However, to date little research has taken place as to whether it is in fact pheromones (ie scent - water is an exceptionally good medium for the transmission of aromas, much better than air) which are the actual attractants. Whatever the answer, it is evident that some males attract ripe females. Because of the type of substrate over which they live, the rock-dwelling mbuna are not particularly inventive in the construction of their spawning sites. On the contrary, the piece of rock that represents the territory is also the spawning site. The male plays the dominant role in the courtship. And, particularly in shallow areas, his behaviour exposes him to the danger of being spotted and captured by aquatic birds. Once a passing female has chosen a male she allows him to lead her to the centre of his territory. The male repeatedly displays his anal fin , on which there are a number of egg dummies in the form of yellow eggspots. He spreads his anal fin on the spawning site, in contact with the substrate, with his body simultaneously at a slight angle. The female now snaps at the egg dummies. After a while the

Gebieten, von Wasservögeln erspäht und erbeutet zu werden. Hat sich ein Weibchen ein passendes Männchen ausgewählt, läßt es sich von ihm zum Zentrum seines Reviers locken. Immer wieder präsentiert das Männchen seine Afterflosse, auf der ein oder mehrere Eiatrappen in Form gelber Eiflecke liegen. Auf dem Ablaichplatz wird die Afterflosse des Männchens in Kontakt mit dem Substrat darauf abgespreizt, wobei der Körper eine leichte Schräglage einnimmt. Das Weibchen schnappt nun nach den Eiatrappen. Anschließend nimmt das Weibchen eine ähnliche Lage wie das Männchen zuvor ein und streift mit dem Bauch über das Substrat. Dabei stupst das Männchen mitunter die Analregion bzw. auch den Schwanzstiel des Weibchens an. Nach einigen Scheinpaarungen kommt es zur Abgabe der Geschlechtsprodukte, die vom Weibchen sofort ins Maul genommen werden (die eigenen Eier, die den Aufnehmreiz darstellen und die Samenzellen des Männchens, die versehentlich beim Schnappen nach den Eiflecken auf der Afterflosse des Männchens aufgenommen werden - wozu die Atrappen schließlich da sind). Dieser Vorgang wiederholt sich mehrfach, wobei jeweils nur wenige Eier pro "Durchgang" gelegt werden.

Im See ist gelegentlich ein spezielles Verhalten zu beobachten, das als "Arena-Brüten" bezeichnet wird. Bekannt ist es z.B. von *Ps. livingstonii* "likoma", *Ps.* sp. "burrower", *M. vermivorus* und *Ps.* sp. "polit". Dabei besetzen mehrere Männchen aneinandergrenzende Reviere und rivalisieren untereinander um die Weibchen. Hierbei treten ihre Körperfarben und bestimmte Zeichnungsmerkmale besonders intensiv hervor. Laichreife Weibchen suchen solche Arenen auf und entscheiden sich hier für einen oder auch mehrere Partner zum Ablaichen. Die Brut eines Weibchens kann also durchaus von meheren Männchen befruchtet sein. Für weitere Informationen zu diesem hochinteressanten Thema s. bitte Fryer & Iles (1972), Ribbink et. al. (1983) und Konings (1995), um nur einige zu nennen.

Hat das Weibchen ausgelaicht, zieht es sich in ruhige Gefilde zurück oder sucht Schutz in Weibchenschulen, wo es zwischen 16 Tagen und etwas mehr als drei Wochen Eier und Larven im Maul allein bebrütet. Durch gelegentliches Umschichten und Versorgen mit Frischwasser erhalten die Eier und Larven ausreichend Sauerstoff. Es wurde schon beobachtet, daß ein brütendes Weibchen nach mehreren Tagen vorsichtig Futter aufgenommen hat, doch scheint dies nicht die Regel zu sein. Sie überstehen die Brutzeit auch ohne Nahrungsaufnahme. Bei Mbunas ist es eher selten, daß die Mütter die Jungen auch nach dem ersten Freischwimmen noch weiter betreuen. In der Regel sind sie fortan auf sich gestellt.

Wir danken dem South African Journal of Zoology für die Erlaubnis, Abbildungen aus Ribbink, A.J., B. A. Marsh, A. C. Ribbink & B. J. Sharp (1983): A preliminary survey of the cichlid fishes of rocky habitats in Lake Malawi, für diese Einleitung übernehmen zu dürfen.

female adopts the same position as the male and brushes her belly across the substrate while the male, meanwhile, nudges her anal region and/or caudal peduncle. After a few "dummy runs" eggs are produced and immediately taken into the mouth of the female (not only the eggs themselves, which are the immediate object of her attentions, but also the milt of the male, which is unwittingly ingested as she snaps at the eggspots on the anal fin of the male - which is what they are there for). This process is repeated, with a few eggs being laid at each "pass".

It has been observed in the lake that a group of male animals occupy adjacent territories and compete with each other for females; this is called "arena breeding" and known, for example, from *Ps. livingstonii* "likoma", *Ps.* sp. "burrower", *M. vermivorus*, and *Ps.* sp. "polit". The competition appears to enhance their coloration and the vigour of their courtship display. Females visit the arena when ready to mate, and may choose a single partner or breed with several different ones. For further information on this highly interesting theme please see Fryer & Iles (1972), Ribbink et. al. (1983) and Konings (1995), just to name some well known sources.

When the female has finished spawning she retires to a quiet spot or seeks out the safety of a shoal of females, where she alone broods eggs and fry for between 16 days and more than 3 weeks. The eggs and larvae are periodically "churned" and provided with fresh water, and thus receive sufficient oxygen. It has been observed that a brooding female may sometimes carefully take food after several days, but this appears not to be the rule - normally she completes the brooding period without feeding. It is very rare for mbuna to continue to tend their brood after first release. As a rule the fry are immediately left to their own devices.

We like to thank the South African Journal of Zoology for their permission to use figures from Ribbink, A.J., B. A. Marsh, A. C. Ribbink & B. J. Sharp (1983): A preliminary survey of the cichlid fishes of rocky habitats in Lake Malawi for this introduction.

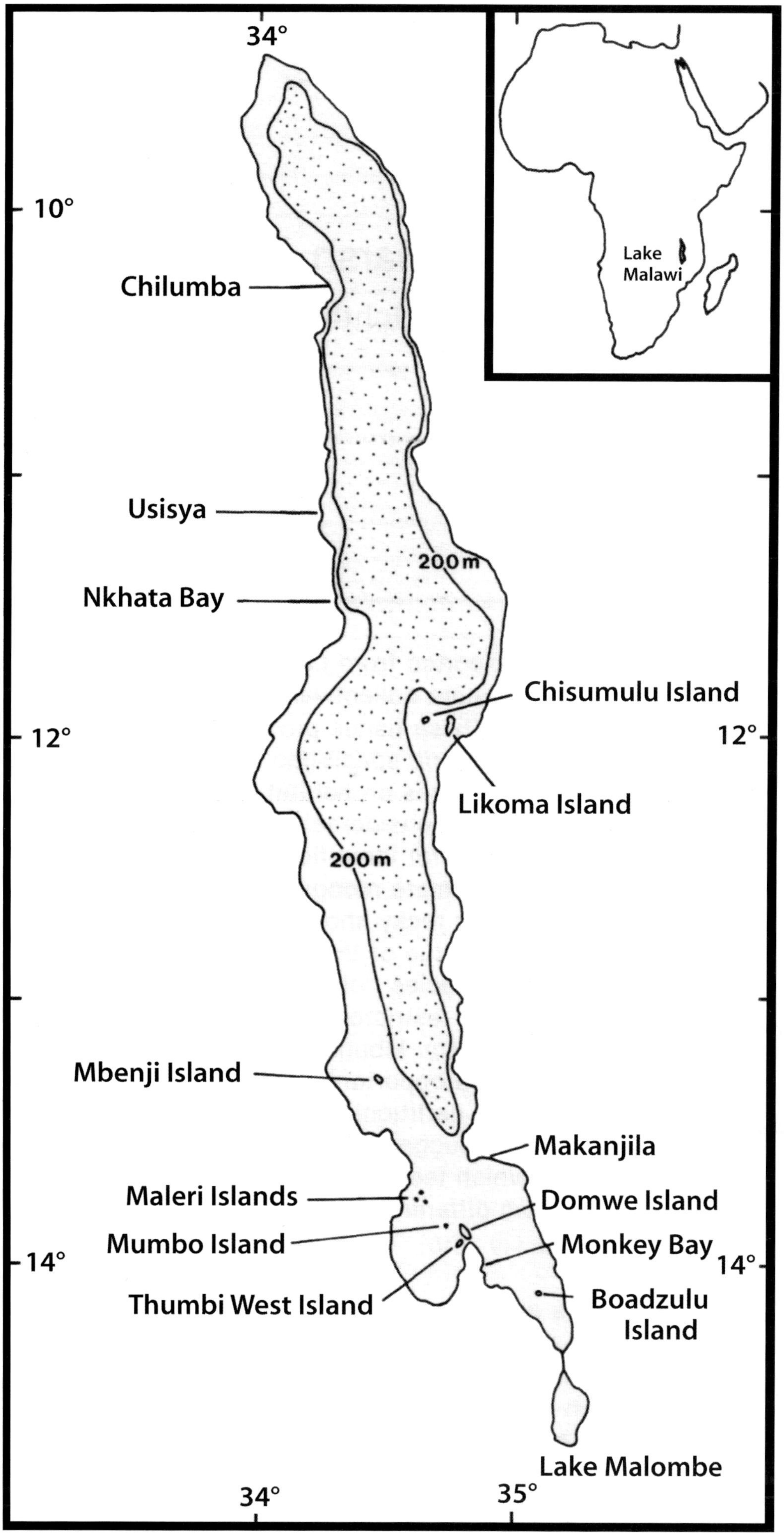

Eine Übersichtskarte des Malawisees. Das eingeklinkte kleine Bild zeigt die Lage des Sees in Afrika. Die Zone, in der das Wasser Tiefen von 200 m und mehr erreicht, ist gesondert gekennzeichnet.

A map of Lake Malawi with an insert of Africa showing the location of the lake. The position of the 200 m depth contoure is indicated.

from: Ribbink et. al.,1983

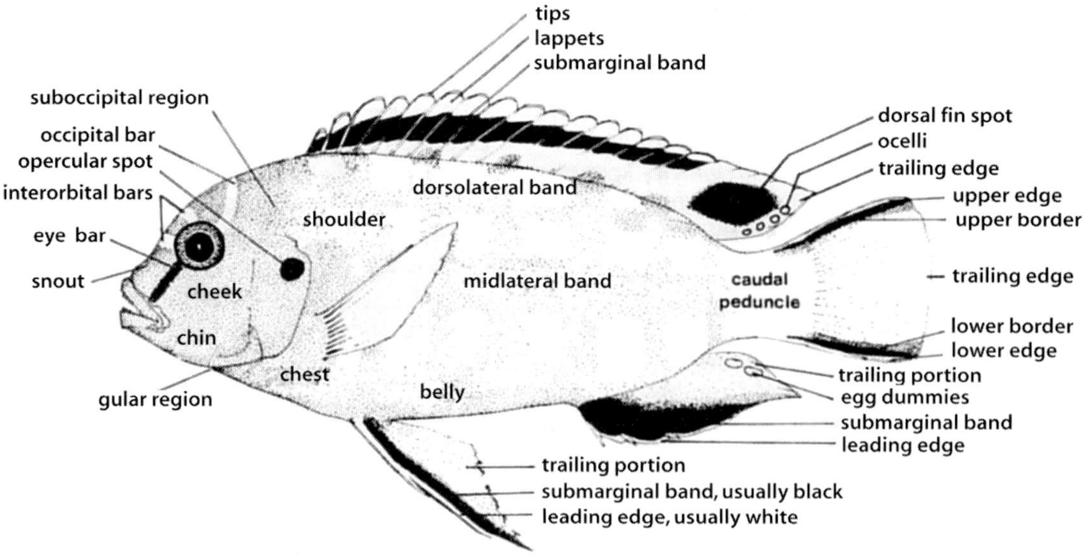

Darstellung der typischen Zeichnungselemente eines Mbunas und ihrer international üblichen Bezeichnung.
An illustration of the markings and the terminology used to describe them.

from: Ribbink et. al., 1983

Bemerkungen zu der Revision der Arten um *Pseudotropheus zebra*

von Ad Konings

Im November 1997 veröffentlichten STAUFFER, BOWERS, KELLOGG und MCKAYE eine Arbeit mit dem vielversprechenden Titel: "A revision of the blue-black *Pseudotropheus zebra*-complex from lake Malawi, Africa, with a description of a new genus and ten new species" (Eine Revision der Blau-Schwarzen Arten des Pseudotropheus-zebra Komplexes aus dem Malawisee, Afrika, mit der Beschreibung einer neuen Gattung und zehn neuer Arten). Dr. Jay STAUFFER und seine Mitarbeiter gehören zu den wenigen Arbeitsgruppen, die sich zur Zeit mit der Taxonomie der Malawicichliden befassen. Während der letzen zehn Jahre beschrieben sie 22 neue Arten aus dem See, eine erstaunlich hohe Zahl. Und jetzt kommen noch einmal zehn Arten hinzu! Abgesehen von der Beschreibung der neuen Arten kommt noch die Beschreibung einer neuen Gattung, nämlich *Metriaclima*, hinzu, zu der jetzt alle schwarz gestreiften Arten des *Pseudotropheus zebra* Komplexes (im Sinne von RIBBINK et. al.) zählen. Ich bin froh, daß sich Dr. STAUFFER dieses seit langem überfälligen Projektes angenommen hat.

Cichlidenfreunde erinnern sich vielleicht noch daran, daß 1984 MEYER & FOERSTER bereits eine (Unter-) Gattung für diesen Artenkomplex errichtet hatten, nämlich *Maylandia* mit *Pseudotropheus geshakei* als Typusart. STAUFFER und Mitarbeiter kamen jedoch zu der Überzeugung, daß es sich bei dieser Beschreibung nicht um eine gültige Gattungsbeschreibung handele, weil die Gattungsdefinition nicht ausreiche, um nachfolgenden Wissenschaftlern die Einordung von Arten in diese Gattung zu ermöglichen. Sie

Comments on the Revision of *Pseudotropheus zebra*

by Ad Konings

In November 1997 STAUFFER, BOWERS, KELLOGG, and MCKAYE published a paper with the interesting title: "A revision of the blue-black *Pseudotropheus zebra*-complex from lake Malawi, Africa, with a description of a new genus and ten new species". Dr. Jay STAUFFER and his coworkers are one of the few groups currently working on Malawi cichlid taxonomy and have been so for the past 10 years. During this period they described 22 new Malawi species, a staggering number. And now 10 new species more! Apart from the species descriptions also a new genus is described, *Metriaclima*, and all the black-barred species previously grouped in RIBBINK et al.'s "*Ps. zebra*-complex" are now grouped in this new genus. I'm glad that Dr. STAUFFER and his coworkers undertook this vast project as it was long overdue.

Cichlid specialists with a good memory may remember that in 1984 MEYER & FOERSTER proposed already a name for the species in the complex, they coined the name *Maylandia* and *Ps. greshakei* was nominated as the type species of the genus. However, when STAUFFER and his colleagues read and reread MEYER & FOERSTER'S paper they came to the conclusion that there was no description of the new genus — no characters were given that future taxonomist could use in sorting their species — and thus it follows that the name *Maylandia* is a nomen

erklärten *Maylandia* daher zum nomen nudum und ersetzten den Namen durch *Metriaclima,* obwohl *Maylandia* an sich durchaus weiterhin verfügbar gewesen wäre.

Metriaclima bedeutet soviel wie "sanft ansteigend", was sich auf die charakteristische Kopfform dieser Tiere bezieht. Diese Kopfform kommt von der gattungstypischen Ausprägung des sogenannten etmo-vomerinen Blocks, eines nasenartigen Auswuchses amSchädel (genauer: die rostrale Erweiterung des Neurocraniums, die den Oberkieferapparat unterstützt), der einen etwa 45°-Winkel zur Körperlängsachse bildet, wogegen der etmovomerine Block bei dem *P. tropeops*-Komplex einen 90°-Winkel zu Körperlängsachse bildet. Ferner wird auf den Schädel von *P. elongatus* verwiesen, bei dem diese Struktur ebenfalls einen Winkel größer 45° bildet. Leider wird nicht erwähnt, welche der zahlreichen Arten des *P. elongatus*-Komplexes untersucht wurde, also ob es sich um den Holotypen oder bloß ein Mitglied des Komplexes handelte. Es ist schon länger unter Liebhabern bekannt, daß es zwei Typen von Cichliden innerhalb dieses Komplexes gibt: den einen, dessen Kopfform an *P. zebra* erinnert und einen zweiten, der eher Ähnlichkeit zu *P. tropheops* aufweist. Mir scheint es so, als ob die Röntgenaufnahme einen Fisch mit tropheops-ähnlichem Schädel zeigt. Wenn das der Fall wäre, könnte man annehmen, daß die zebra-ähnlichen Formen des *P. elongatus*-Komplexes auch einen an *P. zebra* erinnernden etmo-vomerinen Block hätten. Das würde weiterhin dazu führen, daß diese schlanken Arten zu *Metriaclima* zu zählen wären, zumindest nach der bisherigen Definition von *Metriaclima.* Dem könnte ich durchaus zustimmen, doch ein anderer Taxonom könnte sich da anders entscheiden und das würde wiederum eine Neudefinition von *Metriaclima* notwendig machen.

Ich glaube, daß grundsätzlich das Splitting der Gattung *Pseudotropheus* auf breite Akzeptanz stoßen wird und daß, bis die Sache mit den zebra-artigen Fischen des *P. elongatus*-Komplexes geklärt ist, der Name *Metriclima* für die schwarz gestreiften Mbuna verwendet werden sollte, die zuvor im *P. zebra*-Komplex zusammengefasst waren. Ohne größeren Umstand können wir außerdem ab sofort den Namen *Tropheops* als Gattungsnamen für die Arten verwenden, die Ribbink et. al. im *P. tropheops*-Komplex zusammenfasste. *Tropheops* war von Trewavas 1984 als Untergattung zu *Pseudotropheus* für eben diesen Komplex vorgeschlagen und aufgestellt worden und kann ohne weitere Beschreibung in vollen Gattungsrang erhoben werden. Übrigens erschien im gleichen Journal im gleichen Heft die Arbeit von Meyer und Foerster, in der die Autoren eine Untergattung für den *P. zebra*-Komplex vorschlagen. Aufgrund der Defintion von *Tropheops* sind in dieser Gattung derzeit die tropheops-artigen Fische des *P. elongatus*-Komplexes eingeschlossen , weil in der Defintion von

nudum. So quite right Stauffer et al. suggested another name (they could have used the same name because a nomen nudum is still available for new descriptions): *Metriaclima.*

The name *Metriaclima*, meaning "moderately sloped to indicate the moderately sloped head of members of this genus", alludes to the fact that all species investigated had a so-called ethmo-vomerine block (the nose-like extension of the skull — the rostral extension of the neurocranium that supports the upper jaw) which makes about a 45 degree angle with the body axis whereas species of the *Ps. tropheops*-complex have a ethmo-vomerine block that makes an almost 90 degrees angle with the body axis. They also refer to the species of the *Ps. elongatus*-complex in which the ethmo-vomerine block also makes a larger angle. Unfortunately they don't mention which of the many species in the *Ps. elongatus*-complex they investigated, i.e. whether it was the holotype or just a member of the complex. It has been known to me and scores of other cichlid enthusiasts that in the *Ps. elongatus*-complex there are two different types of elongated cichlids: one with heads resembling that of *Ps. zebra* and one resembling that of *Ps. tropheops.* It seems to me that they show an X-ray photo of a *tropheops*-type elongated mbuna and if this is the case then we could anticipate that the *zebra*-type species of the *Ps. elongatus*- complex have an ethmo-vomerine block which resembles that of *Ps. zebra.* This would further lead to the conclusion that those elongated species also belong to *Metriaclima* as currently defined. I could agree to that but if a later taxonomist decides that those should reside in another genus he or she would have to redefine *Metriaclima* as well.

I think that in general the splitting up of *Pseudotropheus* will find acceptance and that, until the *Ps. elongatus* complex issue is resolved, *Metriaclima* is used for the black-barred mbuna species previously grouped in the *Ps. zebra* complex. Without further much ado we can also start using the generic name *Tropheops* for the species currently in Ribbink et al.'s *Ps. tropheops* complex. *Tropheops* was suggested and defined as a subgenus by Trewavas in 1984 for these species and can without further much ado be elevated to full generic status. As a matter of fact her description preceded the Meyer and Foerster paper in the same journal and she mentions that the latter two authors propose a new subgenus for the zebras. As the definition is at the moment the *Tropheops*-

Tropheops nicht auf den Index Körperhöhe zu Körperlänge eingegangen wird.

STAUFFER et. al. grenzen *Metriaclima* des weiteren gegen *P. williamsi*, die Typusart von *Pseudotropheus*, ab. Bevor STAUFFER et. al. ihre neuen Arten beschreiben, liefern sie eine Wiederbeschreibung von *Metriaclima zebra*. Sie haben den Holotypen (dessen exakte Herkunft unbekannt ist) nachuntersucht und sind dabei zu der Erkenntnis gelangt, daß dieser Holotyp in Details nicht mit all dem anderen Material von *M. zebra*, das sie untersuchen konnten, übereinstimmt. Die Unterschiede beziehen sich allerdings auf Größenordnungen im Bereich von 1 bis 2 Zehntel eines Millimeters. In Anbetracht der Tatsache, daß der Holotyp über hundert Jahre in Alkohol gelegen hat, verwundert das nicht allzusehr. Was mich jedoch beim Studium der Maßtabellen überraschte, war die Tatsache, daß dieser Holotyp wesentlich größer als alle anderen vermessenen *M. zebra* war. Üblicherweise schrumpfen Fische in Alkohol. Davon, daß Fische in Alkohol wachsen, habe ich zumindest noch nicht gehört. In der Diskussion ihre Arbeit sagen STAUFFER et. al., daß sie ursprünglich vorhatten, den Artnamen *M. zebra* auf den Holotypen zu beschränken. Aufgrund der weiten Verbreitung dieses Namens bei Wissenschaftlern und Liebhabern beschlossen sie dann aber doch, unter dieser Artbezeichnung alle bisher bekannten und übereinstimmend als *M. zebra*-Populationen akzeptierten Fische zu führen und den Holotypus als atypisches Exemplar zu definieren. Das ist ein Wiederspruch in sich und wenn wir die Liste der aufgeführten zebra-Populationen betrachten, so stellen wir fest, daß sie bei weitem nicht alle bekannten Populationen umfasst. Die meisten der untersuchten Exemplare kommen aus dem südöstlichen Arm des Sees und einige aus der Umgebung der Nkhata Bay. Interessanterweise beziehen sie Exemplare von Chitande Rocks ein. Wenn es sich dabei um einen Ort nahe Chilumba handelt (dort liegt Chitande Island und es könnte sich um einen Schreib-/Verständigungsfehler bez. Mitande Rocks handeln), dann gibt es ein Problem, denn dort kommt *M. zebra* nicht vor. Ich glaube, sie haben den sog. Chilumba Zebra untersucht (weil sie eine gelbe Kehlregion und einen schwarzen Streifen in der Rückenflosse bei nördlichen Populationen erwähnen), der von den meisten Cichliden-Experten nicht als *M. zebra* angesehen wird. Meiner Ansicht nach ist der Chilumba Zebra näher mit *M. faintzilberi* verwandt (der von den Autoren nicht zu *Metriaclima* gerechnet wird) und bevorzugt wie *M. faintzilberi* Habitate, die sich deutlich von denen unseres *M. zebra* unterscheiden.

Als ich die ersten Gerüchte darüber, daß STAUFFER den Namen *P. zebra* auf den Holotypen beschränken und damit zu einer Museumsleiche machen wollte, zu Ohren bekam, dachte ich, das wäre etwas übertrieben und daß das so niemand akzeptieren würde. Jetzt, da ich die Tabellen studiere und annehmen

type species of the *Ps. elongatus* complex are included, because in the diagnosis of *Tropheops* nothing is mentioned about the body depth/standard length ratio.

STAUFFER et al. give further characterization of *Metriaclima* to exclude *Ps. williamsi*, the type species of *Pseudotropheus*. Before STAUFFER et al. discuss the new species they give a redescription of *Metriaclima zebra*. They have investigated the holotype, which is of unknown origin, and found that the holotype does not fit in the morphology of all the other *M. zebra* material they have investigated. The discrepancies are, however, in the order of magnitude of 1 to 2 tenth of a millimeter. And given the fact that the holotype has been in alcohol for more than 100 years nobody would be surprised to find that the fish doesn't have the exact same measurements as when it was caught. However, when I look at their tables it strikes me that the holotype is much bigger than all other zebras measured. Fish generally shrink in alcohol but of growing fish I haven't heard. In their discussion they mention that initially they wanted to retain the name *M. zebra* to the holotype solely, but on grounds that this name is in "wide usage in both the scientific and popular literature" they include all the populations we currently know and regard as *zebra* and mention that the holotype is "atypical". This is a contradiction in terms and if we look at the list of *zebra* specimens they have investigated we see that, by far, they have not covered all populations; most of the specimens originate from the southeast arm of the lake and some from Nkhata Bay area. Interestingly they include specimens from Chitande Rocks. If this locality is near Chilumba (Chitande Island and there might have been a typographic mix up with Mitande Rocks) then we have a problem here because *M. zebra* has not yet been found there. I think they have investigated the so-called Chilumba Zebra (since they mention yellow gullar and a black stripe in the dorsal in northern populations), which certainly is not regarded as a *M. zebra* by most cichlid specialists. In my opinion the Chilumba Zebra is closely related to *M. fainzilberi* (which is NOT regarded as belonging to *Metriaclima*! by the authors) and shares with it its habitat preference which differs clearly from that of our *M. zebra*.

When I first heard rumors that STAUFFER was going to restrict the name *Ps. zebra* to the holotype only and thus make a museum specimen out of it, I thought that this was a little overdoing it and expected that

muß, daß sie (wahrscheinlich aus anatomischen Gründen) den Chilumba Zebra zu ihrem *M. zebra* zählen, fürchte ich, es wäre besser gewesen, sie wären bei ihrer ursprünglichen Absicht geblieben und hätten den Namen *M. zebra* auf den Holotypen und vielleicht noch die größeren zebra-Typen (inklusive des Chilumba Zebras) beschränkt. So besteht jetzt aber nach wie vor große Unsicherheit bezüglich des Holotypen. Denn hätten STAUFFER et. al. Exemplare von Likoma untersucht und mit dem Holotypen verglichen, so wüßte ich jetzt wenigstens, welche Meinung ich mir bilden sollte. Der zebraartige Fisch von Likoma wurde schon immer zu *P. zebra* gezählt, obwohl bekannt ist, daß er größer wird als alle mir bekannten *P. zebra*-Populationen (vielleicht mit Ausnahme der von Taiwan Reef) und daß er definitv mehr Zahnreihen hat. Letzteres ist genau, was STAUFFER et. al. beim Holotypen von *P. zebra* gefunden haben: 5 Zahnreihen im Unterkiefer, im Vergleich zu 2 bis 4 Zahnreihen bei allen anderen Zebras der Art *M. zebra* (sensu STAUFFER et. al.). Das Titelbild meines Buches über Malawi-Cichliden aus dem Jahre 1989 (roter Umschlag) zeigt eine Frontalaufnahme eines Likoma Zebras. Man kann hier beinahe die Zahnreihen durchzählen: das Männchen hat mindestens 5! Wenn (auch wenn wir das wahrscheinlich niemals mit Sicherheit erfahren werden) der Holotyp von Likoma Island stammt, und das ist gar nicht so unwahrscheinlich, weil die Insel seinerzeit ein religiöses Zentrum war und dort Briten siedelten, dann sind sind wir nach wie vor mit der Frage konfrontiert, ob diese Fische mit all den anderen Zebras artgleich sind. Wenn es sich aber herausstellt, daß wir nicht in der Lage sind, zwischen den großen Zebras vom *M. faintzilberi*-Typ und den "ordentlichen" schwarz-blauen Zebras zu unterscheiden, sind wir immer noch weit davon entfernt, die großen Zebras von Likoma einer der beiden Gruppen sauber zuzuordnen.

An vielen anderen Orten (besonders im tansanischen Bereich des Sees) werden große Mbuna vom *M. fainzilberi*-Typ sympatrisch mit *M. zebra* gefunden. Bei Mphanga Rocks, nur wenige Meilen von Chitande Island entfernt, finden wir Chilumba Zebras zusammen mit Red Top Zebras, einer Art, die viel enger mit *M. zebra* verwandt ist (und die ich nach wie vor als geographische Variante des letzteren ansehe). Die Morphologie und die Bevozugung bestimmter Habitate des Red Top Zebra ist sehr ähnlich der von *M. zebra*. Jemand, der diese Cichliden praktisch überall im See beobachtet hat, kann und wird niemals akzeptieren, daß der große Chilumba Zebra zur gleichen Art wie die schwarz-blauen Zebras, die überall sonst gefunden werden, gehört. Unglücklicherweise ist daher das letzte Wort über den Holotypen und die Artbezeichnung von *M. zebra* noch nicht gesprochen.

Die neue Art *M. melabranchion* unterscheidet sich von *M. zebra* dadurch, daß bei ihr die senkrechten

almost nobody would accept that. Now that I have the tables in front of me and the fact that they, probably on bases of morphology, include the Chilumba Zebra in their *M. zebra*, I fear that they should have pursued their initial goal and restrict the name *zebra* to the larger holotype and perhaps include the large *zebra* types, including the Chilumba Zebra. Now we are still uncertain about the holotype because if STAUFFER et al. would have investigated specimens from Likoma Island and compared those with the holotype, at least I would have known what to decide. The Zebra type found at Likoma Island has always been regarded as *Ps. zebra* although it is known to be larger than all those of other *zebra* populations I know (perhaps with the exception of those at Taiwan Reef) and it has definitively more rows of teeth in their jaws. The latter is exactly what STAUFFER et al. found in the holotype: 5 rows of teeth in the lower jaw whereas all other Zebras in *M. zebra* have 2 to 4. The frontispiece of my book on Malawi cichlids published in 1989 (red cover) shows a frontal shot of a Likoma zebra and the tooth rows can almost be counted (the male has at least 5 rows!). If, and this we will probably never know, the holotype stems from Likoma Island (which is not unreasonable as the island was the religious centre in those days with residing British nationals) then we still are facing the problem of including it with the other zebra cichlids. If we prove to be unable to distinguish between the large *M. fainzilberi* type zebras and the "regular" BB zebra we are still far from deciding whether the large *zebra* at Likoma Island belongs to either one of them.

At many other sites large *M. fainzilberi* type mbuna are found sympatrically with *M. zebra*, especially in the Tanzanian part of the lake. At Mphanga Rocks, just a few miles from Chitande Island, we find the Chilumba Zebra together with the Red-Top Zebra, a species much more related to *M. zebra* (and still regarded by me as a geographical variant of it). The morphology and habitat preference of the Red-Top Zebra are much more like those of *M. zebra*. Someone who has studied these cichlids in their natural habitat at almost all sites around the lake will and can never accept that the large Chilumba Zebra is the same as the BB zebra found elsewhere. Unfortunately the last word on the holotype and the name *M. zebra* has not been spoken.

The new species *M. melabranchion* differs from *M. zebra* in the fact that the vertical bars on the body are

schwarzen Körperstreifen sich bis in die Rückenflosse hinein ausdehnen. Die Autoren stellen ferner fest, daß die Art bei Mitande Rocks sympatrisch mit *M. zebra* lebt, was als weiteres Kriterium dafür herangezogen wird, daß es sich um unterschiedliche Arten handelt. Wenn das zutrifft, dann lebt nahezu jede Population sympatrisch mit *M. zebra*, denn ich habe Fotos von Namalenje Island und Domwe Island (beides Verbreitungsgebiet von *M. melabranchion*), die *M. zebra* ohne dieses Merkmal zeigen. Der Vergleich der in der Tabelle aufgeführten Merkmale zeigt keinerlei morphologische Unterschiede zu *M. zebra* und, was interessanter ist, keinerlei Hinweis auf irgendwelche arttrennenden Mechanismen, die ein Vermischen der beiden Arten bei Mitande oder anderen Orten, wo sie sympatrisch vorkommen, verhindern könnten (dergleichen hätte die Argumente der Autoren entscheidend untermauert). Wenn die Ausprägung der senkrechten Streifung das einzige arttrennende Merkmal wäre (es ist zumindest das einzige angegebene), dann wäre es interessant zu erfahren, wie jemand nur daran festmachen kann, daß es sich um zwei unterschiedlich genetische Pools handelt und nicht nur um einen polymorphen Faktor, so wie die OB-Morphen, die man in den meisten Populationen findet. Über Zuchtexperimente und eine Untersuchung der Jungen ist nichts bekannt. Dr. Irv Kornfield und ich haben große Mengen von Zebras im Feld untersucht und gefunden, daß in allen Populationen Männchen mit und Männchen ohne diese Ausdehnung der Streifung in die Rückenflosse vorkommen. Das kommt sogar beim Red Top Zebra vor (und ich habe Fotos, die das beweisen!). Eine ähnliche Situation findet sich bei verschiedenen anderen Mbuna-Arten, so z.B. bei *Cynotilapia*- und *Tropheops*-Arten. Meiner Ansicht nach liegt dieses Merkmal in der Varianz der Art und spielt auch keine Rolle (oder zumindest nur eine sehr kleine) in der Erkennung der Männchen durch die Weibchen. Zumindest scheint mir ein flächiger Streifen in der Rückenflosse in dieser Hinsicht viel entscheidender zu sein. Bemerkswert ist, daß die Chilumba Zebras, die einen solchen Streifen aufweisen, von den Autoren zur gleichen Art gehörig angesehen werden, wie die *M. zebra* mit eine ungezeichnete, einheitlich hellblauen Rückenflosse. Diese Tatsache wird in der Arbeit nicht weiter kommentiert und auf S. 215 widersprechen sich die Autoren selbst, indem sie feststellen, daß es keine Zeichnungselemente innerhalb *M. zebra* gäbe, abgesehen von der blau-schwarzen Streifung.

Die Beschreibung von *M. chrysomallos* erstaunt mich insofern, als daß das einzige Unterscheidungsmerkmal zu *M. melabranchion* die goldene Sprenkelung auf hellblauer Grundfärbung sein soll. Die Art ist endemisch bei Mumbo Island. Obwohl ich oft bei Mumbo Island getaucht habe, habe ich nie bemerkt, daß es dort zwei unterscheidbare Zebras gibt. Ich dachte, die helleren Exemplare schwämmen üblicherweise über sedimentreicherem Bodengrund,

extended into the dorsal fin. The authors further note that "the fact that *M. melabranchion* is sympatric with *M. zebra* at Mitande Rocks supports the conclusion that these two taxa are heterospecific." If this were correct then almost every population of *M. melabranchion* is sympatric with *M. zebra*, because I have photos taken at Namalenje Island and Domwe Island showing *M. zebra* without the bars extended into the dorsal fin. Comparison of the tabled data for *M. melabranchion* with that for *M. zebra* shows no morphological differences and, more interestingly, NO evidence of any character displacement for sympatric populations at Mitande or other localities (something that would have made their argument more meritorious). If the extended bars is the only distinction (it is the only one given) then it would be interesting to know how one can find out whether these two forms indeed belong to two separate genetic pools or that the bar extension is just a polymorphic factor, such as OB morphs, found in most populations. Breeding experiments and the analysis of the offspring was not reported. Dr. Irv Kornfield and I have been observing the dorsal fins of lots of zebra cichlids in the wild and clearly found that in ALL populations there are males with and males without bar extensions, even in Red-Top Zebras (and I have photos to prove this). A similar situation is found in several other mbuna species, e.g. *Cynotilapia* and *Tropheops* species. In my opinion those bar extensions are part of the variability of the species and do not play a role (or a very minor one) in mate recognition, at least not to the same extent as a solid stripe in the dorsal does. Remarkably the Chilumba Zebra which has a very distinct stripe in the dorsal fin is regarded by the same authors as conspecific with the non-marked, light-blue dorsal-finned *M. zebra*. This peculiarity is not further mentioned in the paper and on page 215 they contradict themselves by mentioning that there are "no other pigmentation patterns" apart from the blue/black color pattern in *M. zebra*.

The description of *M. chrysomallos* puzzles me as its only distinction from *M. melabranchion* is "the golden highlights on the light blue ground coloration". It is endemic to Mumbo Island. Although I have dived many times at Mumbo Island I was never aware that there are two distinguishable zebras present. I thought that the light-colored individuals were usually found in more sediment-rich areas whereas the darker forms more in pure rocky environments. This I have seen all over the lake but maybe we are

wogegen die dunkleren Tiere eher über reinem Steinboden zu finden seien. Das habe ich im ganzen See gesehen, aber vielleicht sprechen wir ja über verschiedene Arten. Ich muß aber festhalten, daß die angegebenen Artunterschiede wirklich minimal und an konserviertem nicht nachvollziehbar Material sind.

Die nächste Art, die die Autoren beschreiben, ist *M. phaeos*. Der Holotyp stammt von Cobue, Mosambik. Die Autoren stellen fest, daß *M. phaeos* oberflächlich *M. zebra* sehr ähnelt, aber daß das wolkige, submarginale Band in der Rückenflosse die Art klar von allen anderen blau-schwarzen Vertetern der Gattung unterscheidet. In der gleichen Arbeit (auf S. 193) schreiben sie jedoch, daß nördliche Populationen von *M. zebra* einen durchgängigen , waagerechten schwarzen Streifen in der Rückenflosse hätten. Obwohl ich zustimme, daß *M. phaeos* eine von *M. zebra* unterschiedliche Art ist (zumindest wenn es sich dabei um die Art handelt, an die ich dabei denke), kann zumindest nach Färbungsmerkmalen eine sichere Artabgrenzung nicht vorgenommen werden. Ich habe 5 mal bei Cobue getaucht und weil es sich um ein überschaubares Gebiet handelt, glaube ich ganz gut zu wissen, was dort für Mbuna-Arten schwimmen. Der einzige blau-schwarze Mbuna bei Cobue ist die Art, die gemeinhin als "Shauri" bekannt ist. Es gibt auch einen *M. zebra* bei Cobue (meine Determination), aber der zeigt keine Streifung. Demnach glaube ich, die Beschreibung der Männchenfärbung von *M. phaeos* meint den "Shauri", aber die Beschreibung der Weibchenfärbung paßt so gar nicht zu den gelben Shauri-Weibchen. Die Weibchen von *M. phaeos* sollen dunkel blaugrau sein und ein Fünftel der unteren Flankenhälfte gelb-braun. Der Rest der Farbbeschreibung spricht von gelbbraunen Farbtönen in allen anderen Körperpartien und so könnte es sich bei der Angabe "dunkel blaugrau" um einen Satzfehler handeln oder die Weibchen wurden mit der ebenfalls dort vorkommenden Art des *M. aurora*-Typs verwechselt.

Der Holotyp von *M. cyneusmarginatus* (siehe zu der Schreibweise cyneus - cyaneus die Bemerkung von Michael Oliver im Internet: http://www.connix. com/ ~mko/) wurde bei Nkhotakota in sehr seichtem Wasser gefangen. Sein blaues marginales Band und die braunen "Lappets" (gemeint sind die nach oben verlängerten Teile der Rückenflossenmembran. Anm. d. Übers.) unterscheidet die Art von anderen *Metriaclima*. Ich habe noch nie Felsen im seichten Wasser bei Nkhotakota gesehen, aber einige Meilen südlich von dieser Stadt, nahe bei dem Zugang zur Chia Lagune, fand ich einen Mbuna, den ich Zebra Benga genannt habe und auf den die Beschreibung von *M. cyneusmarginatus* paßt.

Die Beschreibung von *M. benetos* beruht auf der Tatsache, daß in sedimentreichen Gebieten die Zebra-Cichliden üblicherweise einen wesentlich helleren blauen Farbton als in felsigen Habitaten und

dealing with different species. I must say that the distinction given is very minimal and in preserved specimens this is not retraceable.

The next species described is *M. phaeos* and the holotype originates from Cobue, Mozambique. The authors mention that "*M. phaeos* superficially resembles *M. zebra* but that the dusky submarginal band in the dorsal fin of *M. phaeos* distinguishes it from all other BB members of the genus." In the same paper (p. 193), however, they mention that some of the northern populations of *M. zebra* have a "continuous black horizontal strip" in the dorsal fin. Although I agree with the authors that *M. phaeos* is distinct from *M. zebra*, at least as it is what I think it is, a valid distinction in coloration has not been given. I have dived about 5 times at Cobue and as the site is very small I think to know more or less what mbuna are present. The only dark blue black barred mbuna I know at Cobue is the species commonly known as the "Shauri". There is a *M. zebra* at Cobue (my identification) but it shows no barring. So, the description of the male colors of *M. phaeos* match those of the Shauri male, but the female colors do not match those of the yellow Shauri females. *M. phaeos* females are "dark blue/gray with ventral one-fifth of flank yellow/brown". The rest of the color description mentions yellow/brown on all other parts and the "dark blue/gray" may in fact be a typographical error (or there might have been a mix up with females of the *M. aurora*-type species present at the same locality).

The holotype of *M. cyneusmarginatus* (see note on spelling **cyneus** vs. **cyaneus** by Michael Oliver at http://www.connix.com/~mko/) was caught at Nkhotakota in very shallow water. "The blue marginal band and brown lappets distinguishes it from other members of *Metriaclima*". I have never found rocks in shallow water at Nkhotakota but a few miles south of the town, near the entrance of Chia Lagoon, I found a mbuna which I have called Zebra Benga and which match the description of *M. cyneusmarginatus*.

The description of *M. benetos* is based on the fact that in sediment-rich areas zebra cichlids usually have a much lighter blue color than in pure rocky habitats, and often males do not show the vertical bars. Mazinzi Reef in the southeast arm of the lake is such a place but also Jalo Reef, Membe Island south of Chizumulu, Chiwindi, Undu Reef, and Cobue. "The pale blue ground color with the absence of distinct

daß die Männchen oft ihre senkrechte Streifung nicht zeigen. Mazinzi Reef im südöstlichen Arm des Sees ist ein solcher Ort, doch findet man Vergleichbares auch bei Jalo Reef, Membe Island südlich von Chizumulu, Chiwindi, Undo Reef und Cobue. Die blasse blaue Grundfärbung und das Fehlen von ausgeprägten senkrechten Binden unterscheidet *M. benetos* von allen anderen Vertretern der Gattung mit Ausnahme von *M. callainos*. Nach Aussage der Autoren ist *M. benetos* endemisch am Mazinzi Reef. Auf die hellblauen Zebras bei Cobue wird nicht weiter eingegangen, obwohl sie den Fisch beim Sammeln von *M. phaeos* (dem Shauri) gesehen haben müssen, denn er ist ziemlich häufig. Es ist Ansichtssache, ob man den hellblauen Zebra als artverschieden von dem schwarz-blauen Zebra betrachtet. Aber in Anbetracht der Tatsache, daß Bilder von hellblauen Zebras von den oben aufgeführten Fundorten vor 2 Jahren, von einigen sogar schon vor 14 Jahren, publiziert wurden, wäre es schön gewesen, wenn die Autoren sich dazu geäußert hätten, ob es sich dabei auch um *M. benetos* handelt oder nicht. Zum besonderen Ärgernis für die "Splitters" unter uns: es gibt auch Übergangsformen. Wie sollten diese eingeordnet werden?

In der Wiederbeschreibung von *M. callainos* (dem Cobalt Zebra) wird darauf hingewiesen, daß die Art künstlich bei Likoma, Namalenje und Thumbi West Islands angesiedelt wurde. Ich habe die Art nie bei Namalenje gefunden und es könnte sich bei dieser Angabe um einen Satzfehler handeln und eigentlich Maleri Island gemeint sein. Die Autoren bemerken ferner, daß der Pearl Zebra von der Nordwestküste des Sees nicht zu *M. callainos* gehört und eine Artbeschreibung in Vorbereitung ist. Der Unterschied soll in der weißen Färbung beider Geschlechter liegen. Allerdings schwimmen bei Katale Island kobalt-blaue Fische. Ich hoffe, diese Population wird bei einer zukünftigen Arbeit nicht übersehen.

Die Autoren fanden eine weitere Population von *M. xanstomachus* bei Kanjedza Island im südöstlichen Teil des Sees, womit die Art nun nicht mehr als endemisch für Maleri Island gilt. Nebenbei bemerkt: ich habe sie außerdem bei Chidunga Rocks gefunden.

Die Autoren splitten die Red Top Zebras in 4 verschiedene Arten auf! Von drei dieser Arten wird gesagt, sie hätten eine rote Rückenflosse, von einer, sie hätte eine orangene Rückenflosse. Ich habe alle 4 Populationen selbst untersucht und fand, daß die rote Pigmentation in der Rückenflosse der Männchen sehr variiert und meiner Meinung nach eine Unterscheidung zwischen rot oder orange nicht möglich ist. Sie fallen alle in die Kategorie orange bis rotorange. Abgesehen von der Färbung, die sich bei konservierten Tieren ohnehin verliert, stellen die Autoren ferner fest, daß der Red Top Zebra vom Eccles Reef ein gelbes Kinn hat, was ein zusätzliches

vertical bars distinguishes *M. benetos* from all other members of this genus, with the exception of *M. callainos*". According to the authors *M. benetos* is endemic to Mazinzi Reef. There is no further mention of the pale blue zebra at Cobue which they must have seen while collecting *M. phaeos* (the Shauri), as it is quite common. It is a personal matter to regard the light blue zebra distinct from the BB zebra, but since pictures of the pale blue zebras at the aforementioned localities have been published 2 and some even 14 years ago it would have been better to at least mention whether these are regarded *M. benetos* or not. Unfortunately for the splitters among us there are also intermediate forms; how to classify those?

The redescription of *M. callainos* (the Cobalt Zebra) mentions that it has been introduced at Likoma, Namalenje, and Thumbi West Islands. I have never found it at Namalenje and this may be a typographical error and should be Maleri Island. The authors further mention that the Pearl Zebra from the northwest coast of the lake is distinct from *M. callainos* and is in the process of being described. Their distinction being all white males and females. At Katale Island, however, there are cobalt blue colored individuals. If there is going to be a split-up again then hopefully this population will not be forgotten.

The authors found another population of *M. xanstomachus* at Kanjedza Island in the southeastern arm of the lake and the species is thus not longer endemic to the Maleri Islands. As an aside, I also found it at Chidunga Rocks.

The authors suggest to split the Red-Top Zebras into 4 different species! Three of them are regarded to have a red dorsal fin and one an orange dorsal fin. I have observed all these four populations and found that the amount of red pigment in the dorsal varies among the males and in my opinion there is no such distinction as red or orange. They all fall in the same category of orange to red-orange. Apart from this color, which is lost in preserved fish, they further mention that the Red-Top Zebra from Eccles Reef has a yellow chin which further distinguishes it from the other Red-Tops. And again this feature is found in almost all populations of any zebra, not only Red-Tops. *M. pyrsonotos* is the Red-Top from Nakantenga Island and it is claimed that "the vertical black lateral bars infiltrate the proximal portion of the dorsal fin." I

Unterscheidungsmerkmal zu den übrigen Red Tops sein soll. Und wieder ist das ein Merkmal, das in nahezu allen Populationen irgendwelcher Zebras auftaucht, nicht nur bei den Red Tops. *M. pyrsonotos* ist der Red Top von Nakantenga Island und es wird behauptet, daß sich die schwarzen, senkrechten Bänder bis in die Rückenflosse fortsetzen. Ich habe Fotos von männlichen Red Tops von Nakantenga, wo bei einigen Tieren die Bänder fast bis an den oberen Rand der Rückenflosse reichen und andere, bei denen die Bänder bereits kurz unterhalb des Ansatzes der Rückenflosse enden, diese also nicht einmal berühren. Ferner besitze ich Fotos von Fischen, deren Zeichnung zwischen diesen beiden Extremen liegt. Alle anderen morphologischen Daten, die zu den übrigen Red Tops geliefert werden, überschneiden sich mit denen von *M. pyrsonotos*, so daß keine wirkliche Unterscheidung zwischen den Arten getroffen werden kann. Ich wüßte gerne, was die Veranlassung dafür war, diese Populationen in so viele Arten aufzusplitten. In Anbetracht der Tatsache, daß *M. pyrsonotos* der erste der in der Revision benutzte Name für die Red Tops ist, könnte er für diese Fische benutzt werden, wenn man sie denn als artverschieden von *M. zebra* ansehen will.

Die Erstbeschreibung des Red Top Cobalt-Zebra von Mbenji Island als *M. mbenjii* ist wohl die erste dieser Revision, die von den Cichliden-Experten auf breiter Basis akzetiert werden dürfte, obwohl auch die diese Art bezeichnendenden Merkmale sich mit denen von *M. greshakei* (einem weiteren hellblauen Zebra mit orange-roter Rückenflosse) überschneiden. *M. mbenjii* ist passenderweise auf die Gewässer um Mbenji Island beschränkt und es gibt in der Nachbarschaft keine Populationen von Zebras, die dieser Art ähnlich sind. Es ist schön, nun einen Namen für diese Fische zu haben.

Eine kritische Überprüfung der Daten, die Stauffer et. al. geliefert haben, macht es äußerst wahrscheinlich, daß nur drei der insgesamt zehn neu beschriebenen Arten valide sind: *M. phaeos*, *M. cyneusmarginatus* und *M. mbenjii*. Sie zeigte des weiteren, daß die Neubearbeitung der Zebras, von der wir dachten, sie brächte uns die Klärung vieler offener Fragen, uns eher verwirrt.

Jegliche an mich gerichtete Kommentare sind sehr willkommen. Ich danke Dr. Irv Kornfield und Martin Geerts für ihre wertvollen Kommentare.

have photos in which males Red-Top at Nakantenga have either bars running almost to the edge of the dorsal or where the bars stop short of the upper edge of the flank and do not penetrate the dorsal at all. I also have photos showing intermediates between these two extremes. As all other morphological data given overlaps with those of *M. pyrsonotos* there is not a valid distinction between these 4 species. I would like to know what the reasoning is to split these populations up in so many species. Since *M. pyrsonotos* is the first mentioned in the revision one may use this name for all Red-Top Zebras, if one accepts that it is distinct from *M. zebra*.

The description of the Red-Top Cobalt Zebra from MbenJI Island is probably the first that will be accepted by cichlid specialists even though there are only overlapping characters given in which this species, now named *M. mbenjii*, differs from *M. greshakei* (another light-blue colored Zebra with an orange-red dorsal). *M. mbenjii* is conveniently restricted to Mbenji Island and neighboring populations of Zebra-type cichlids do not resemble it. It is nice to have a name for it.

A critical evaluation of the data presented by Stauffer et al. strongly suggests that only three of the ten described species are valid: *M. phaeos*, *M. cyneusmarginatus*, and *M. mbenjii*. It further shows that the classification of the zebra, of which we thought was straight forward, appears to surprise us.

Any comments on mine are greatly appreciated. I thank Dr. Irv Kornfield and Martin Geerts for their valuable comments.

Eine Artdefinition für Cichliden des Malawisees

KAREN A. KELLOGG
Graduate Program in Organismic and Evolutionary Biology
University of Massachusetts Amherst
Amherst, MA 01003

JAY R. STAUFFER, Jr.
School of Forest Resources
Penn State University
University Park, PA 16802

Vielleicht ist die erste Frage, die man sich in diesem Zussammenhang stellen muß, folgende: Gibt es so etwas wie Arten überhaupt? Mit anderen Worten: Bilden Arten eine natürliche Einheit oder stellen sie nur ein künstliches, der menschlichen Bequemlichkeit dienendes Konstrukt dar? Die persönliche Einstellung eines jeden zu dieser Frage wird dessen Artdefinition stark beeinflussen. Wissenschaftler wie DARWIN (1859), GILMOUR (1940), EHRLICH und HOLM (1963) und HALDANE (1956) glaubten, daß das Artkonzept erfunden wurde, um Ordnung in das natürliche System zu bringen; demnach wäre die Art eine Kategorie, die als Hilfsmittel für die Klassifizierung dienen sollte (SLOBODCHIKOFF 1976). Im Gegensatz hierzu glaubten Autoren wie SIMPSON (1961), MAYR (1963), HENNIG (1966), GHISELIN (1969), DOBZHANSKY (1970), GRANT (1971), HULL (1976) und WHITE (1978), daß Arten wirklich existieren und daß eine Art eine natürliche Grundeinheit bildet. Wir glauben bestimmt, daß Arten eine natürliche Einheit bilden.

Dabei haben zwei grundsätzliche Beobachtungen unsere Einstellung beeinflußt: Erstens werden schon seit langer Zeit einheimische Tiere von den dort lebenden Menschen erkannt und benannt, ohne daß sie sich mit philosophischen Gedanken bezüglich eines Artkonzeptes herumschlagen. Das allein würde noch nicht ausreichen, um die Debatte um die Frage, ob Arten wirklich existieren, zu beenden. Denn möglicherweise haben die Eingeborenen diese Kategorien auch nur erfunden, um über bestimmte Tiere kommunizieren zu können, genau wie das die Wissenschaftler heute tun. Was jedoch aus dieser Tatsache geschlossen werden kann, ist die evolutionäre Konstanz vieler Arten: Die langanhaltende, unveränderte Bezeichnung bestimmter Tiere über viele menschliche Generationen hinweg kann kaum anders interpretiert werden. Zweitens zeigt zumindest die überwiegende Mehrheit der natürlich zusammen vorkommenden (also nicht künstlich zusammengebrachten) Taxa ein artspezifisches Fortpflanzungsverhalten, wodurch Hybridisation nur relativ selten vorkommt. Diese Organismen müssen demzufolge fähig sein, Artgenossen auch als solche zu erkennen und bilden demzufolge eine natürliche

A Species Definition for Lake Malawi Cichlids

KAREN A. KELLOGG
Graduate Program in Organismic and Evolutionary Biology
University of Massachusetts Amherst
Amherst, MA 01003

JAY R. STAUFFER, Jr.
School of Forest Resources
Penn State University
University Park, PA 16802

Perhaps the first question that must be addressed in a discussion regarding the definition of a species is whether species are real. In other words, do species form a natural group or do they represent an artificial category constructed for human convenience? The philosophy one holds of the reality of species will obviously influence how one would define a species. Scholars such as DARWIN (1859), GILMOUR (1940), EHRLICH and HOLM (1963), and HALDANE (1956) believed that the concept of a species was a construct invented so that order could be put into the natural system; thus, the species is a category invented to aid in classification (SLOBODCHIKOFF 1976). Conversely, authors including SIMPSON (1961), MAYR (1963), HENNIG (1966), GHISELIN (1969), DOBZHANSKY (1970), GRANT (1971), HULL (1976), and WHITE (1978) believed that species are real and recognized as cohesive units in nature. We believe that species are indeed natural, and two general lines of evidence have influenced our thinking.

First, there has been a long history of recognizing and naming native fauna by indigenous peoples who were not privy to philosophical arguments regarding species concepts. This does not perhaps lend directly to resolving the debate regarding whether or not species are real, because after all, it would behoove indigenous people to have categories and names for the sake of discussion just as it does scientists today. What can be gained from this perspective, however, is the evolutionary cohesiveness of many species: the long-standing maintenance of isolated taxa as passed through numerous human generations. Secondly, at least the majority of taxa that reside naturally (i.e., not introduced) in sympatry demonstrate species-specific breeding, with relatively few hybridization events. These organisms are certainly able to "recognize" species boundaries and therefore seem to exhibit natural groupings. In addition, if allopatric taxa can be brought into the laboratory

Gruppe. Hinzu kommt noch, daß, werden allopatrische Taxa unter Laborbedingungen zusammengebracht unter denen sie ihren Geschlechtspartner frei wählen können, sie in den weitaus meisten Fällen einen Partner aus der eigenen Gruppe wählen (KELLOGG et al. 1997).

Nun, da wir gezeigt haben, daß Arten wirklich existieren,können wir die schwierige Aufgabe angehen, den Artbegriff zu definieren. Es gibt eine gewaltige Zahl von kontroversen Ansichten zu diesem Thema und wenn man bedenkt, daß allein 21 verschiedene Artkonzepte bislang publiziert wurden (MAYDEN 1997), kann man leicht erkennen, daß wir von einem Konsens diesbezüglich weit entfernt sind. Die Buntbarsche des Malawisees bilden sicherlich ein großes Forum für die Diskussion der verschiedenen Artkonzepte, was an der explosionartigen Diversifikation (das bedeutet, stark vereinfacht ausgedrückt, Veränderlichkeit. Anm. d. Übers.) der Fische in diesem System liegt. Die Artdefinition, die wir für die Cichliden des Malawisees vorschlagen, basiert im wesentlichen auf dem evolutionären Artkonzept, doch enthält es auch Elemente des biologischen und des morphologischen Artkonzeptes.

SIMPSON (1951) hat erstmals ein evolutionäres Artkonzept vorgeschlagen, das WILEY (1978) später so definierte (bitte s. für die Originalzitate immer den nebenstehenden Originaltext. Anm. d. Übers.): "... eine Population einer einzelnen, von einer Urform abstammenden, Entwicklungslinie, die ihre Identität gegenüber anderen solchen Entwicklungslinien beibehält und die ihre eigenen evolutionären Tendenzen und historischen Werdegänge besitzt." Das evolutionäre Artkonzept ist demnach eine Artdefinition, die als einende Grundidee den Gedanken vertritt, daß alle Arten eine Entwicklunglinie darstellen, die getrennt ist von allen anderen Entwicklungslinien anderer Arten. Die Art ist also eine natürliche Einheit auf einem unabhängigen evolutionären Entwicklungszweig. Aufgrund der Tatsache, daß unabhängig von der Form der Fortpflanzung genetisches Material von dem (den) Elterntier(en) an die Nachkommen weitergegeben wird, ist dieses Konzept der Entwicklungslinie auf alle Formen der Fortpflanzung anwendbar. In letzter Konsequenz macht dieses Konzept eine große Anzahl Arten denkbar, sowohl ausgestorbene als auch noch existierende. Wir schlagen vor, daß dieses umfassende Artkonzept angewendet werden sollte, um Cichliden des Malawisees zu determinieren.

Der Nachteil des evolutionären Artkonzeptes liegt allerdings klar auf der Hand: es setzt unbedingt eine solide Theorie zur Phylogenese voraus. Eine Theorie zur Phylogenese ist der Versuch einer Darstellung der Entwicklungsstufen einer Gruppe zu untersuchender Organismen, einschließlich der Rekonstruktion des gemeinsamen Vorfahrens und der Summe der Unterschiede (MAYR und ASHLOCK, 1991). Dieser Werdegang

where they can be given simultaneous choices of potential mates, they most frequently choose a mate of their own group (KELLOGG et al. 1997).

Now that we have addressed the reality of species, we can undertake the arduous task of defining a species. There is a tremendous amount of controversy regarding this topic and with 21 different species concepts appearing in the literature (MAYDEN 1997), there is obviously no consensus. The cichlids of Lake Malawi certainly provide a tremendous forum for the discussion of species concepts, given the explosive diversification within this system. The species definition that we propose for Lake Malawi cichlids is based primarily on the evolutionary species concept, but we also include criteria put forth as part of the biological and morphological species concepts.

SIMPSON (1951) originally proposed an evolutionary species concept, which WILEY (1978) later defined as "...a single lineage of ancestor-descendant populations, which maintains its identity from other such lineages and which has its own evolutionary tendencies and historical fate." The evolutionary species concept is therefore a species definition, which uses as a unifying framework the idea that all species form a lineage, which is separate from all other species lineages. The species is therefore a natural entity on an independent evolutionary trajectory. Since regardless of the type of reproduction, genetic material is passed from parent(s) to offspring, this lineage concept applies to all reproductive modes. In so doing this concept encompasses a wide array of possible species, both extant as well as extinct. Thus, Mayden (1997) concluded that the evolutionary species concept was the only concept that was apropos to all organisms. We propose that this is the overall concept by which species of Lake Malawi cichlids should be determined.

The shortcoming of the evolutionary species concept, however, is that it relies heavily on a good phylogeny. A phylogeny is the inferred lines of descent of a group of organisms including a reconstruction of the common ancestor and the amount of divergence (Mayr and Ashlock 1991). This evolutionary history of a group is typically depicted in a tree diagram. The methods used to construct phylogenies are of course as controversial as species concepts themselves and are certainly beyond the scope of this manuscript. The reality is that in many systems, including Lake Malawi cichlids, the lack of a comprehensive, well supported, phylogeny is

einer Gruppe wird üblicherweise in einem Stammbaum dargestellt. Die Methoden, um zu einer Theorie zur Phylogenese zu kommen, werden freilich genauso kontrovers diskutiert wie die Artkonzepte und ihre Darstellung würden den Rahmen dieses Manuskriptes sprengen. Tatsache ist, daß in vielen Systemen, einschließlich dem der Cichliden des Malawisees, das Fehlen einer vergleichbaren, gut abgesicherten Theorie zur Phylogenese große Probleme bereitet. Daher schlagen wir hier verschiedene Kriterien vor, die bei der Entscheidung über den Artstatus helfen. Das erste ist das der reproduktiven Isolation. Dieses Kriterium ist traditionell mit dem biologischen Artkonzept verknüpft, das MAYR und ASHLOCK (1991) so definierten: " Eine in Hinsicht auf die Fortpflanzung isolierte Gemeinschaft von Populationen, die sich miteinander fortpflanzen können, da sie die gleichen isolierenden Merkmale gemeinsam haben." Reproduktive Isolation kann durch das Beobachten von Paarungaktivtitäten oder die Analyse des Genflusses festgestellt werden.

Wir wollen ganz deutlich feststellen, daß wir die reproduktive Isolation lediglich als Indikator für die Unabhängigkeit einer evolutionären Entwicklungslinie und nicht als Artkonzept per se sehen. Zweitens können wir morphologische und Verhaltenskriterien heranziehen, um Arten abzugrenzen, beides Kriterien, die üblicherweise mit dem morphologischen Artkonzept assoziiert sind. Auch hier müssen wir wieder darauf hinweisen, daß wir nicht das morphologische Artkonzept als solches fördern wollen, sondern nur glauben, daß bestimmte morphologische Unterschiede einen Hinweis auf separate Evolutionslinien liefern. Wie vielleicht bereits deutlich geworden ist, können die beiden genannten Kriterien (also die reproduktive Isolation und die morphologische Differenzierbarkeit) nicht nur als Hilfmittel zur Artabgrenzung herangezogen werden sondern auch zur phylogenetischen Rekonstruktion eines Stammbaumes. Auf den ersten Blick mag dieser Argumentationsgang vielleich doppelt-gemoppelt erscheinen, wir denken aber eher, daß diese Vorgehensweise dazu dient, den bislang klaffenden Spalt zwischen Taxonomie auf alpha-level (also der Arterkennung) und der phylogenetischen Rekonstruktion des Stammbaumes zu überbrücken.

Zunächst müssen hypothetische Arten aufgestellt werden, deren Status dann mit phylogenetischen Methoden überprüft werden muß. Anschließend, wenn man aus phylogenetischer Sicht brauchbare Ansatzpunkte gefunden hat, können diese wiederum dazu verwendet werden, eine gesicherte Artdetermination vorzunehmen. Im Falle der Buntbarsche des Malawisees sind wir noch im Stadium der Hypothesenbildung, obwohl wir schon auf einige vorläufige Phylogenien zurückgreifen können (z. B. PARKER und KORNFIELD 1997). Wir publizieren Arten, die wir aufgrund dessen, was wir über die Zeichnung, die reproduktive Isolation, anatomischen Merkmale

problematic. Hence, we are proposing several criteria to aid in the definition of a species. The first is reproductive isolation. This criterion has been traditionally associated with the biological species concept which MAYR and ASHLOCK (1991) defined as "A reproductively isolated aggregate of populations, which can interbreed with one another because they share the same isolating mechanisms." Reproductive isolation can be assessed through observations of mating activities or analysis of gene flow. We would like to stress that we are simply using reproductive isolation as an indicator of the independence of evolutionary lineages and not as a species concept in and of itself. Secondly, we can employ the morphological/behavioral similarity criteria that is typically associated with the morphological/phenetic species concept to aid in species delineation. Again, we are not promoting the morphological/phenetic species concepts as such, but rather that sufficient morphological difference is evidence for separate evolutionary lineages.

As is perhaps readily evident, the two criteria (i.e., reproductive isolation and morphological differentiation) outlined above can be used in not only species delineation, but also phylogenetic reconstruction. Initially this appears as perhaps circular in reasoning, but we would argue that it points more towards the balance that must be struck between alpha-level taxonomy (species delineation) and phylogentic reconstruction. There must be hypothesized species put forth that are then tested using phylogentic resolution, and then in turn, the phyologenetic resolution will help to adjust our species designations. In terms of Lake Malawi cichlids, we are currently still at the hypothesizing stage, although we are now able to site some preliminary phylogenies (e.g., PARKER and KORNFIELD 1997). We are putting forth species delineations based on what we know about the evolutionary patterns, reproductive isolation, and morphological/behavioral similarities of Lake Malawi cichlids. With the advent of greater phyologenetic resolution through the use of more molecular characters, we may well reject some of the species descriptions made to date. The potential rejection of some species descriptions does not make the overall effort unwarranted. After all, the postulation and rejection of hypotheses is the very nature of science.

The final major issue that must be addressed is allopatric populations. Perhaps DARWIN (1859) was correct in saying that when dealing with allopatric populations "the opinion of naturalists having sound

und Verhaltensweisen wissen, gegen andere Malawicichliden abgrenzen können. Mit mehr Wissen über die Phylogenie dieser Arten, zum Beispiel durch den verstärkten Gebrauch von Methoden auf molekularer Ebene, kann es durchaus sein, daß wir einige der jetzt beschriebenen Arten wieder einziehen müssen. Die Gefahr, das tun zu müssen, macht die hier beschriebene Vorgehensweise aber nicht ungerechtfertigt. Letztendlich ist das Aufstellen und auch das Verwerfen von Arbeitshypothesen der wahre Charakter wissenschaftlicher Arbeit.

Der letzte große Punkt, auf den eingegangen werden muß, ist die Bewertung allopatrischer (also geographisch voneinander isolierter) Populationen. Vielleicht hatte DARWIN (1859) recht, als er sagte, daß "die Meinung von Wissenschaftlern mit sicherem Urteilsvermögen und großer Erfahrung die einzige Richtschnur sein kann, der man folgen kann", um den Status allopatrischer Populationen zu bewerten. Anstatt solchen, rein subjektiven Entscheidungen zu folgen, kann man aber auch unseren Artdefinitionen zustimmen. Sollten wir es tatsächlich mit allopatrischen Populationen zu tun haben, die sich auf einem unabhängigen evolutionären Entwicklungszweig befinden, so sollten wir auch genug Mut aufbringen, sie als eigenständige Arten zu beschreiben. Sollten diese allopatrischen Populationen zufällig wieder in Kontakt zueinander kommen und sich die zwei Entwicklungslinien wieder vermischen, so ist das nicht zwangsläufig ein Argument gegen unsere frühere Annahme. So haben z.B. SEEHAUSEN et. al. (1997) verschiedene Experimente durchgeführt, in denen sie zeigen konnten, daß, wenn visuelle Signale unmöglich werden (z.B. durch zunehmende Wassertrübung), sich Cichliden des Viktoriasees kreuzen. Darüber hinaus nehmen sie an, daß dies der Grund für den nachweislichen Artenschwund in dem See ist. Dieses Verschmelzen von Entwicklungslinien bedeutet aber nicht, daß es nicht viele verschiedene Arten vor der Eintrübung des Viktoriasees gab. Es zeigt uns nur schlicht und ergreifend, wie dynamisch der Prozeß der Evolution und Artbildung ist.

Zusätzlich zu den allgegenwärtigen entwicklungsgeschichtlich bedingten Zeichnungsmerkmalen können auch unsere Kriterien der eingeschränkten Partnerwahl, der morphologischen Unterschiede und der Verhaltensmuster bis zu einem gewissen Grad zur Bewertung allopatrischer Populationen herangezogen werden. Wenn wir die Mechanismen, die bei sympatrischen (also zusammen vorkommenden) Populationen zu einer selektiven Partnerwahl führen, genau analysieren, erhalten wir damit eine Meßlatte, an der sich der Grad morphologischer Unterschiede und Unterschiede im Verhalten beim Vergleich allopatrischer Populationen messen läßt (MAYR und ASHLOCK 1991). Bei den Malawicichliden hat sich das Zeichnungsmuster als ein wesentlicher Punkt für die innerartliche Arterkennung bei sympatrischen Populationen herausgestellt (HOLZBERG 1978, DEUTSCH 1997;

judgment and wide experience seems the only guide to follow." In lieu of using strictly subjective decisions, however, our proposed species definition can be applied. If indeed we have allopatric populations that are on independent evolutionary trajectories, we should certainly be confident in describing them as separate species. If by chance these allopatric populations were brought back into contact and the two lineages merged that does not necessarily negate our original delineation. For example, SEEHAUSEN et al. (1997) conducted several experiments that demonstrated that when visual cues were obstructed (i.e., by turbidity) in Lake Victoria cichlids, they readily interbreed. Furthermore, they proposed this as a possible mechanism that led to the documented decline in cichlid diversity. This merging of lineages does not mean that we did not have many different species prior to the clouding of Lake Victoria waters, it simply reminds us of the dynamic nature of evolution and speciation.

In addition to the overall evolutionary pattern, our criteria of mating isolation and morphological/behavioral similarity can be applied to a certain degree for allopatric populations. Using sympatric populations that demonstrate assortative mating, we can use the degree of morphological/behavioral difference as a yard stick that can then be used for comparing allopatric populations (MAYR and ASHLOCK 1991). In Lake Malawi cichlids, it appears that coloration patterns are particularly critical for species recognition when in sympatry (HOLZBERG 1978, DEUTSCH 1997; SEEHAUSEN et al. 1998). Likewise, when allopatric populations have been given simultaneous choices of mates, many choose conspecifics based primarily on body color (KELLOGG et al. 1997). In addition, overall shape differences have been quantified for assortatively mating sympatric populations as well as courtship behaviors (STAUFFER et al., 1995). Given that there is empirical evidence for the usefulness of these two characteristics, we have used them heavily in our species descriptions.

Using the species definition outlined above, we recently published a revision of the blue-black *Pseudotropheus zebra* complex (STAUFFER et al., 1995)., and we describe a new genus as well as 10 new species. First, we want to explain why we changed the generic name to *Metriaclima* from *Pseudotropheus. Pseudotropheus* was originally described with *Pseudotropheus williamsi* being designated as the type species. This fixes the generic name of *P. williamsi* as *Pseudotropheus.* Since the specimens that we described are distinct from *P. williamsi* and form a

Seehausen et al. 1998). Gibt man allopatrisch vorkommenden Populationen im Experiment die Möglichkeit zur freien Partnerwahl, so wählen viele, primär aufgrund von Zeichnungsmerkmalen, Partner ihrer Population (Kellogg et al. 1997). Darüber hinaus konnte gezeigt werden, daß das allgemeine Erscheinungsbild bei sympatrischen Populationen sowohl bei der Partnerwahl als auch beim Paarungsverhalten eine wesentliche Rolle spielt (Stauffer et al., 1995). Aufgrund der empirischen Erkenntnis über die Brauchbarkeit dieser beiden Charakteristika zur Artabgrenzung haben wir ihnen in unseren Artbeschreibungen einen großen Stellenwert eingeräumt.

Unter Verwendung der oben genau beschriebenen Artdefinition haben wir kürzlich eine Revision des blau-schwarzen *Pseudotropheus zebra* Komplexes veröffentlicht (Stauffer et al., 1995) und sowohl eine neue Gattung beschrieben als auch zehn neue Arten. Zunächst wollen wir erläutern, warum wir den Namen *Pseudotropheus* für diese Fische nicht beibehalten haben, sondern statt dessen *Metriaclima* wählten. Als Gattungstyp für *Pseudotropheus* wurde die Art *Ps. williamsi* festgelegt, womit der Name nur für *P. williamsi* und für Fische, die eng mit *P. williamsi* verwandt sind, zur Verfügung steht. Weil aber die von uns bearbeiteten Arten deutlich von *P. williamsi* abweichen und eine unabhängige Verwandtschaftsgruppe darstellen, mußten sie in Übereinstimmung mit dem International Code of Zoological Nomenclature in eine separate Gattung überführt werden. Obwohl schon früher *Maylandia* als Untergattung für verschiedene Mitglieder dieser Gruppe vorgeschlagen wurde, erfolgte auf diesem Vorschlag keine formelle Beschreibung oder Diagnose der Untergattung; Daher erklärten wir *Maylandia* zum nomen nudum (nackter Name ohne Relevanz) und somit handelt es sich bei *Metriaclima* nicht um ein Junior-Synonym (eine jüngere Doppelbeschreibung) .

Metriaclima kommt aus dem Griechischen und bedeutet in etwa "sanft ansteigend", was sich auf die Wölbung des Kopfes dieser Tiere bezieht. Der Name enthält also schon einige Information darüber, was für die Mitglieder der Gattung charakteristisch ist. Wir haben die Schädel von *Pseudotropheus tropheops* und eines noch unbeschriebenen *Pseudotropheus* der *P. elongatus*-Gruppe (einer schlanken Form) von Mazinzi Reef mit dem Schädel von *Metriaclima zebra* verglichen, um die charkteristische Schädelform zu demonstrieren (Stauffer et al. 1997, figs. 1-3, pp. 192-193). Sollten wir innerhalb der *P. elongatus*-Gruppe (Ribbink et al. 1983) tatsächlich Arten finden, die über einen leicht abfallenden Kopf, eine verdickte rostrale Spitze des Neurocraniums und isognathe Kiefer haben, ferner in den äußeren Zahnreihen zweispitzige Zähne aufweisen und andere Charakteristika, wie sie für Mbuna typisch sind, so würden sie, der bisherigen Definition zufolge, zu *Metriaclima* gehören. Sollten aber allgemeine Körpermerkmale deutlich abweichend sein (z.B. schlank gegenüber

cohesive group, they must be placed in a separate genus according the International Code of Zoological Nomenclature. Although *Maylandia* was suggested as a subgenus for certain members in this group, no formal description or diagnosis accompanied this suggestion; thus *Maylandia* is considered to be a nomen nudum (naked name) and is not a senior synonym of *Metriaclima. Metriaclima* from the Greek, means moderately sloped and indicated the moderately sloped head of members of the genus. Therefore, the name itself imparts some knowledge of the characteristics of the genus. We compared the skulls of *Pseudotropheus tropheops* and an undescribed *Pseudotropheus* sp. *elongatus* (an elongate form) from Mazinzi Reef to *Metriaclima zebra* to demonstrate the moderate slope of the head (Stauffer et al. 1997, figs. 1-3, pp. 192-193). If indeed we find species that were originally included in the "elongatus" group (Ribbink et al. 1983) that possess a moderately sloped head, swollen rostral tip of the neurocranium, isognathus jaws, and have bicuspid teeth in the outer rows of the jaws, as well as the other characteristics that define mbuna, then they too would belong in *Metriaclima* as currently diagnosed. If overall body shapes are extremely different (i.e., elongate versus robust), however, it might appropriate to refine and sharpen the description of the genus *Metriaclima* and generate a new genus for the elongate forms.

In terms of the species that were described in the aforementioned paper, we recognize that there are many populations of these forms throughout Lake Malawi, and this work was the first attempt to diagnose these species. To think that all of the descriptions put forth in Stauffer et al. (1997) will go unchallenged and unmodified in the future would not only deny scientific progress, but also the dynamic nature of evolution and speciation within the lake. The descriptions simply represent what we believe to be the most accurate conclusions under the guise of the evolutionary species concept, with the data we have to date. We will briefly discuss the more controversial of our species descriptions in the context of our species definition. With respect to *Metriaclima zebra*, there was a clinal grade in the overall shape and color patterns of the allopatric populations from south to north. Certainly if you compared a northern population with a southern one, you would have complete isolation. In fact in an earlier paper (Stauffer et al. 1995), one of the authors (J.R.S.) hypothesized that the Nkhata Bay and Thumbi West populations of the blue-black color form represent two distinct species. Upon further analysis,

hochrückig), könnte es nötig werden, die Gattungsdiagnose von *Metriaclima* zu präzisieren und eine neue Gattung für die schlanken Formen aufzustellen.

Bezüglich der Arten, die wir beschrieben haben, ist uns sehr wohl klar, daß es noch viele Populationen dieses Artenkreises im ganzen See gibt und die oben genannte Arbeit war nur ein erster Versuch, Richtlinien für die Diagnose solcher Arten zu erstellen. Anzunehmen, alle Arten, die in STAUFFER et al. (1997) beschrieben wurden, würden in Zukunft ohne Wiederspruch und ohne Veränderungen bestehen bleiben, würde nicht nur dem wissenschaftlichen Fortschritt wiederspechen, sondern auch der dynamischen Natur der Evolution und Artbildung im See. Diese Beschreibung stellen lediglich das Ergebnis dessen dar, was wir mit größter Sorgfalt aus den uns derzeit vorliegenden Daten unter dem Aspekt des evolutionären Artkonzeptes herausgefiltert haben und zu wissen glauben. Wir wollen hier einmal kurz auf die umstritterenen Arten im Kontext mit unserer Artdefintion eingehen.

Bezüglich *Metriaclima zebra* fanden wir heraus, daß es bezüglich der allgemeinen Körperform und des Zeichnungsmusters eine klinale Veränderung bei den allopatrischen Populationen vom Süden zum Norden hin gibt. Sicherlich, würde man eine nördliche Population mit einer südlichen vergleichen, so fände man eine komplette Isolation. Und tatsächlich hat einer der beiden Autoren (J.R.S.) in einer früheren Arbeit (STAUFFER et al. 1995) die Hypothese vertreten, die Populationen der blau-schwarzen Zebras von Nkhata Bay und Thumbi West würden unterschiedliche Arten darstellen. Weitere Untersuchungen zeigten dann aber, daß, bewegt man sich vom südlichen Teil des Sees in den Norden, eine Population fließend in die nächste übergeht und daß somit eine vollständige Überlappung aller Merkmale festzustellen war (STAUFFER et al. 1997). Derzeit werden Untersuchungen angestellt, um die genetische Struktur von felsbewohnenden Population in einem sehr feine geographischen Raster zu entschlüsseln (MARKERT et al., in prep, DANELY et al., in prep). Die vorläufigen Ergebnisse dieser Untersuchungen zeigen, daß isolierte genetische Strukturen bereits in überraschend geringen geographischen Abständen zueinander existieren. Sollten sich diese Befunde innerhalb dieses feinen geographischen Rasters bestätigen, so würde das bedeuten, daß Populationen, die in unserer Arbeit noch als *M. zebra* geführt werden, tatsächlich eigenständige evolutionäre Entwicklungszweige darstellen und später als eigenständige Arten beschrieben werden müssen. Unter diesem Aspekt könnte man auch die etwas abweichende Population von Chitande Rocks, die Konings (in diesem Buch) hervorhebt, betrachtet werden, bei der die oberen Strahlen der Rückenflosse orange gefärbt sind, während ein durchgängiger horizontaler schwarzer Streifen durch die Flosse läuft. Wir haben sie zu *M. zebra* gerechnet. Rückblickend, however, we observed that as you moved from the southern part of Lake Malawi to the north, one population graded into another, so that you had complete overlap (STAUFFER et al. 1997). Currently, however, there is much work being done in terms of analyzing the genetic structure of rock-dwelling populations along a very fine geographic scale (e.g., MARKERT et al., in prep, DANELY et al., in prep). Preliminary results of these investigations show isolated genetic structure along surprisingly small distances. If indeed isolation is found on this fine scale, the populations of *M. zebra* described in our paper may well be defined as separate species at a later date given that they represent independent evolutionary lineages. Along this same vein, KONINGS (in this publication) points out a somewhat aberrant population from Chitande Rocks that possesses a dorsal fin with tips of posterior rays orange and a continuous black horizontal stripe which we included in the species *M. zebra*. In retrospect, based on our species definition and the distinct coloration pattern of this population, perhaps we should have described this as a different species. We hesitated doing this in our recent publication because there was significant overlap in the shape of this and the other *M. zebra* populations. Given that we were only able to infer evolutionary independence based on one character, we wanted to reserve judgment of the specific status of this population until additional observational data could be collected. Finally, KONINGS (in this publication) questions the exclusion of *Pseudotropheus fainzilberi* in *Metriaclima*. As stated in STAUFFER et al. (1997), the placement of *P. fainzilberi* STAECK in *Metriaclima* is problematic. Although its body coloration superficially resembles *M. zebra*, it differs primarily because of its peculiar mouth shape and teeth position, which more closely resembles that of *Petrotilapia* (STAECK 1976, fig. 12, p. 489).

In terms of the redtop forms that we examined, the morphological data of the two populations of *Metriaclima sandaracinos* at Kanjedza Island and Mpandi Island were not significantly different in shape and color, and were thus considered conspecific. In addition, given their close proximity we did not believe that these two populations were on independent evolutionary trajectories, although again, genetic data may provide information that will support the delineation of these two populations. The morphological variation of the populations of *Metriaclima emmiltos* and *Metriaclima pyrsonotos* were significantly different from each other and from *Metriaclima sandaracinos*, thus three distinct redtop species were recognized. Similarly, our delineation

legt man unser Artkonzept und die unterschiedlichen Zeichnungsmerkmale zugrunde, hätten wir sie vielleicht als separate Art beschreiben sollen. Wir haben es in unserer Arbeit jedoch deshalb nicht getan, weil es signifikante Überschneidungen in Bezug auf die Körperform dieser und anderer Populationen von Zebras gab. Da wir demzufolge nur ein Merkmal zur Verfügung hatten, das eine evolutionäre Unabhängigkeit dieser Population nahelegte, wollten wir eine endgültige Beurteilung dieser Form noch hinausschieben, bis mehr und besseres Datenmaterial vorliegen würde. Schließlich stellt KONINGS (in diesem Buch) die Ausgrenzung von *Ps. faintzilberi* aus *Metriaclima* in Frage. Wie schon in STAUFFER et.al. (1997) festgestellt wurde, ist die Zuordnung von *P. faintzilberi* STAECK zu *Metriaclima* problematisch. Obwohl die grundsätzlichen Zeichnungsmerkmale sehr stark an *M. zebra* erinnern, unterscheidet sich die Art doch durch ihre einzigartige Maulform und ihr Bezahnungsmuster, das viel stärker an *Petrotilapia* erinnert (STAECK 1976, fig. 12, p. 489).

Bezüglich der von uns untersuchten Red Tops ist zu sagen, daß die Populationen von *Metriaclima sandaracinos* von Kanjedza Island und Mpandi Island so viele Übereinstimmungen bezüglich Körperform und Färbung zeigten, daß wir sie als zu einer Art gehörig betrachten. Hinzu kam noch, daß wir aufgrund des Verbreitungsmusters nicht glaubten, daß es sich bei den zwei Populationen um unabhängige evolutionäre Entwicklungzweige handelt. Aber auch hier müssen wir wieder einschränken: Genetische Untersuchungen könnten eine Trennung der beiden Populationen notwendig machen. Die morphologische Varianz innerhalb *Metriaclima emmiltos, M. prysonotos* und *M. sandaracinos* wich derart signifikant voneinander ab, daß hier drei verschiedene Arten der Red Top-Gruppe angenommen werden müssen. Die Gründe, die gänzlich blau gefärbten Zebras von Nkhata Bay, *M. callainos*, und Mazinzi Reef, *M. benetos*, als verschiedene Arten voneinander zu trennen, waren dem vergleichbar. Auch hier gab es signifikante Unterschiede im Habitus und der Färbung. Zusätzlich zu den Farbunterschieden bei geschlechtsaktiven Männchen soll hier noch erwähnt werden, daß einige Weibchen von *M. callainos* weiß gefärbt sind, was bei *M. benetos* noch nie beobachtet wurde. Wie KONINGS betont, gibt es weitere gänzlich blau gefärbte Zebras an verschiedenen, über den ganzen See verstreuten Fundorten. Nichts desto trotz waren wir uns sicher, aufgrund der Ortstreue, die Cichliden zeigen, des disjunkten Verbreitungsmusters all dieser Populationen und den oben erwähnten Befunden bei der Untersuchung genetischer Strukturen, aus dem uns vorliegenden Material zwei gute Arten beschreiben zu können.

Die Beschreibung von *M. melabranchion* basiert auf der Untersuchung verschiedener Populationen. Die Art unterscheidet sich von *M. zebra* im allgemeinen Habitus und durch Unterschiede der Färbung der

of the all-blue zebra forms at Nkhata Bay, *M. callainos*, and Mazinzi Reef, *M. benetos*, is supported by shape as well as color differences. In addition to differences in breeding male body coloration, some *M. callainos* females are white which has never been observed in M. benetos. As KONINGS points out, there are other all-blue zebra forms at other localities throughout the lake. However, given the disjunct nature of these populations, in conjunction with the philopatric nature of cichlids and the genetic structure that is being identified along fine geographic scales (see above), we were confident in describing the two populations for which we had specimens as separate species.

Metriaclima melabranchion was described based on several populations and differs from *M. zebra* based on overall shape as well as a difference in the dorsal fin coloration; the lateral vertical bars extend into the dorsal fin in *M. melabranchion*. Admittedly, if we found no overall shape differences between these two groups, we would have been hesitant to describe them as separate species based solely on the coloration differences. Given that there were significant shape differences (which can only be assessed using a multivariate statistical analysis and not simply by comparing measurements of different linear measurements), however, we were confident with our designations. In terms of the controversy regarding the ubiquity of bar extensions in *M. zebra* populations, we can only say that in all of our samples of *M. zebra* we did not find bar extensions, and likewise, in all of our samples of *M. melabranchion* we did find bar extensions. Finally, given the fine level of discrimination that we are finding in terms of mate selection, and the function of the dorsal fin in courtship, it is certainly plausible that this small difference in coloration pattern acts as an isolating mechanism.

In closing, we would like to stress that species definitions are not simply an exercise in philosophical conviction. The determination of the specific status of local taxa is critical for the development of programs both to conserve and to utilize these fishes for food, tourism, disease control, trade, and scientific investigation (see STAUFFER et al. 1995 for further discussion). We would assert that conservation programs should focus on evolutionary uniqueness, and hence, independent evolutionary lineages. This position is supported by our proposed species definition, and in turn, our species descriptions are guided by this definition. If we simply described species based on gross overall morphology without

Rückenflosse: Die senkrechten Körperstreifen reichen bei *M. melabranchion* bis in die Rückenflosse. Zugegebenermaßen, hätten wir nicht in allen Populationen habituelle Unterschiede gefunden, hätten wir aufgrund der relativ geringen Farbunterschiede die Arten nicht getrennt. Aber weil eben zu diesen Farbunterschieden signifikante Unterschiede im Habitus festzustellen waren (die man allerdings nur unter Anwendung multivarianter statistischer Methoden und nicht durch einfaches Ausmessen erkennen kann), waren wir uns unserer Sache sicher. Angesichts der erfolgten Widersprüche bezüglich der Ausdehnung der Körpersteifen bei *M. zebra* können wir nur sagen, daß bei unseren Sammlungen von *M. zebra* niemals eine solche Ausdehnung der Streifen auftrat, umgekehrt aber bei allen unseren Exemplaren von *M. melabranchion* diese Ausdehnung vorhanden ist. Schließlich bleibt festzustellen, daß aufgrund der sehr feinen Unterschiede, die bei der Partnerwahl ausschlaggebend sind und der großen Rolle, die die Präsentation der Rückenflosse beim Balzverhalten spielt, es sehr plausibel erscheint, daß selbst geringe Unterschiede bezüglich dieses Färbungsmerkmals eine isolierende Funktion haben.

Abschließend wollen wir noch hervorheben, daß Artdefinitionen nicht nur eine theoretisch-philosophische Stilübung sind. Die Beurteilung des Artstatus von lokalen Populationen ist von großer Bedeutung für die Entwicklung von Programmen, die sowohl dem Schutz wie auch der Nutzung der Fische zu Nahrungszwecken, aber auch durch Tourimus, Krankheitskontrolle, den Handel und Wissenschaft (s. STAUFFER et. al. 1995 für weitere Informationen zu diesem Thema) dienen. Wir wollen darauf hinarbeiten, daß Artenschutzprogramme sich auf die evolutionäre Einzigartigkeit beziehen und demzufolge auf den Schutz unabhängig voneinander existierender evolutionärer Entwicklunglinien. Dieser Standpunkt war ausschlaggebend für unser Artenkonzept und umgekehrt sind unsere Artbeschreibungen unter diesem Aspekt zu betrachten. Wenn wir Arten nur nach einem grob-anatomischen Raster beschrieben, ohne dabei auf den Fundort oder die evolutionäre Einzigartigkeit Rücksicht zu nehmen, würden wir die Diversität der Cichliden in grobster Weise unterschätzen. In einem System wie dem Malawisee, wo die Fische duch Überfischung, die Einfuhr exotischer Arten, die Zerstörung des Lebensraums und die Wasserverschmutzung bedroht sind, ist eine Bestandsaufnahme der tatsächlich vorhandenen Diversität ein dringendes Gebot der Stunde. Wir können die Cichliden des Malawisees nur effektiv schützen, wenn ein der besonderen Situation angepasstes Artenkonzept zur Verfügung steht.

Literatur: wie nebenstehend

(Anm. d. Übers.: Es ist ein heikles Unterfangen, wissenschaftliche Originalarbeiten zu übersetzen. Ich habe mich in der deutschen Version zugunsten einer allgemeinverständlichen Fassung entschlossen, auch wenn sie sich wörtlich (aber, wie ich hoffe, niemals inhaltlich) gelegentlich von der Vorlage entfernt. Ich bitte diesen Punkt zu beachten und bei eventuellen Rückfragen bei den Autoren sich immer nur auf ihren Originaltext zu beziehen. FS)

regard to locality or evolutionary uniqueness, we would be grossly underestimating cichlid diversity. In a system such as Lake Malawi, where fishes are threatened by overfishing, introduction of exotics, habitat destruction, and pollution of the environment, quantifying diversity is a pressing issue. We can only approach the conservation of Lake Malawi cichlids armed with an accurate definition of a species.

Literature Cited

Darwin, C. 1859. The Origin of Species. London.

Deutsch, J. C. 1997. Colour diversification in Malawi cichlids: evidence for adaptation, reinforcement or sexual selection? Biol. Jour. Linn. Soc. 62:1-14.

Dobzhansky, T. 1970. Genetics of the Evolutionary Process. Columbia University Press, New York.

Ehrlich, P. and R. W. **Holm.** 1963. The Process of Evolution. McGraw-Hill, Inc., New York.

Ghiselin, M. T. 1969. The Triumph of the Darwinian Method. University of California Press, Berkely.

Gilmour, J. S. L. 1940. Taxonomy and Philosophy. 461-468. In: Hurley (ed.). The New Systematics. Clarendon Press, London.

Grant, V. 1971. Plant Speciation. Columbia University Press, New York.

Haldane, J. B. S. 1956. Can a species concept be justified? Syst. Assoc. Publ. 2: 95-96.

Hennig, W. 1966. Phylogenetic Systematics. University of Illinois Press, Urbana.

Holzberg, S. 1978. A field and laboratory study of the behaviour and ecology of Pseudotropheus zebra (Boulenger), an endemic cichlid of Lake Malawi (Pisces; Cichlidae). Z. zool. Syst. Evolut. - forsch. 16:171-187.

Hull, D. L. 1976. Are species really individuals? Syst. Zool. 25: 174-191.

Kellogg, K. A. 1997. Lake Malawi cichlid mating systems: factors that influence mate selection. Ph.D. thesis, The Pennsylvania State University, University Park, PA.

Mayden, R. L. 1997. A hierarchy of species concepts: The deouement in the saga of the species problem. In: M. F. Claridge, H. A. Dawah & M. R. Wilson (eds.). The Units of Biodiversity – Species in Practice. Systematics Association, London.

Mayr, E. 1963. Animal Species and Evolution. Harvard University Press, Cambridge, Massachusetts.

Mayr, E. and P. D. **Ashlock.** 1991. Principles of Systematic Zoology (2nd ed.). McGraw-Hill, Inc. New York.

Parker, A. and I. **Kornfield.** 1997. Evolution of the mitochondrial DNA control region in the mbuna (Cichlidae) species flock of Lake Malawi, East Africa. J. Mol. Evol. 45:70-83.

Ribbink, A. J., B. A. **Marsh,** A. C. **Marsh,** A. C. **Ribbink,** and B. J. **Sharp.** 1983. A preliminary survey of the cichlid fishes of rocky habitats in Lake Malawi. S. Afr. J. Zool. 18(3):149-310.

Seehausen, O., J. J. M. **van Alphen.** 1998. The effect of male coloration on female mate choice in closely related Lake Victoria cichlids (Haplochromis nyererei complex). Behav. Ecol. Sociobiol. 42:1-8.

Seehausen, O., J. J. M. **van Alphen,** and F. **Witte.** 1997. Cichlid fish diversity threatened by eutrophication that curbs sexual selection. Science. 277:1808-1811.

Simpson, G. G. 1951. The species concept. Evolution 5: 285-298.

Simpson, G. G. 1961. Principles of Animal Taxonomy. Columbia University Press. New York.

Slobodchikoff, C. N. 1976. Concepts of Species. University of California Press, Berkely.

Staeck, W. 1976. Ergebnisse einer ichtyologischen Sammelreise zum Nordende des Nyassasess. Das Aquarium. 10:486-492.

Stauffer, J. R., Jr., N. J. **Bowers,** K. A. **Kellogg,** and K. R. **McKaye.** 1997. A revision of the blue-black Pseudotropheus zebra (Teleostei: Cichlidae) complex from Lake Malaëi, Africa, with a description of a new genus and ten new species. Proc. Acad. Nat. Sci. Philadelphia 148: 189-230.

Stauffer, J. R., Jr., N. J. **Bowers,** K. R. **McKaye,** and T. D. **Kocher.** 1995. Evolutionarily significant units among cichlid fishes: The role of behavioral studies. Amer. Fish. Soc. Symp. 17: 227-244.

White, M. J. D. 1978. Modes of Speciation. W. H. Freeman and Co. San Francisco, California.

Wiley, E. O. 1978. The evolutionary species concept reconsidered. Syst. Zool. 27: 17-26.

Symbole

Um möglichst alle Fische im Bild zeigen zu können und um dem weltweiten Vertrieb unserer Bildbände Rechnung zu tragen, haben wir bewußt auf ausführliche Texte verzichtet und ersetzen diese durch internationale Symbole, mit deren Hilfe jeder leicht die wichtigsten Eigenschaften der Fische und deren Pflege erkennen kann.

Ursprung:

ersehen Sie ganz leicht an dem Buchstaben vor der Code-Nummer

A = Afrika **E** = Europa + Nordamerika
S = Südamerika **X** = Asien + Australien

Alter:

die letzte Zahl der Code-Nummer steht immer für das Alter des fotografierten Fisches:

1	=	klein	(Jugendfärbung)
2	=	mittelgroß	(Jungfisch / juvenil / Verkaufsgröße)
3	=	groß	(halbwüchsig / gute Verkaufsgröße)
4	=	XL	(ausgewachsen / adult)
5	=	XXL	(Zucht-Tier)
6	=	show	(Schau-Tier)

Herkunft:

W = Wildform
B = Nachzucht
Z = Zuchtform
X = Kreuzungs-Form

Größe:

..cm = ungefähre Größe, die dieser Fisch ausgewachsen (adult) erreichen kann.

Geschlecht:

♂ männlich ♀ weiblich ♂♀ Paar

Temperatur:

18-22°C	(64 - 72°F)	(Zimmertemperatur)
22-25°C	(72 -77°F)	(tropische Fische)
24-29°C	(75 - 85°F)	(Diskus etc.)
10-22°C	(50 - 72°F)	(kalt)

pH-Wert:

pH 6,5 - 7,2 keine besonderen Ansprüche (neutral)
pH 5,8 - 6,5 liebt weiches u. leicht saures Wasser
pH 7,5 - 8,5 liebt hartes u. alkalisches Wasser

Beleuchtung:

○ hell, viel Licht / Sonne
◑ nicht zu hell
◕ fast dunkel

Futter:

Allesfresser, Trockenfutter, keine besonderen Ansprüche
Lebendfutter, Gefrierfutter
Fischräuber, Futterfische füttern
Pflanzenfresser, Pflanzenkost zufüttern

Schwimmverhalten:

keine besonderen Eigenschaften
im oberen Bereich / Oberflächenfisch
im unteren Bereich / Bodenfisch

Aquarium-Einrichtung:

nur Bodengrund und Steine etc.
Steine / Wurzeln / Höhlen
Pflanzen-Aquarium mit Dekoration

Verhalten/Vermehrung:

Paarweise oder im Trio halten
Schwarmfisch, nicht unter 10 Exemplaren halten
Eierleger
Lebendgebärer
Maulbrüter
Höhlenbrüter
Schaumnestbauer
Algenvertilger / Scheibenputzer (Wurzeln+Spinat)
leichte Pflege (für entsprechende Gesellschaftsbecken)
schwierig zu halten, vorher Fachliteratur beachten
Vorsicht, extrem schwierig, nur für erfahrene Spezialisten
die Eier benötigen eine spezielle Behandlung
§ geschützte Art, (WA), "CITES" Sondergenehmigung nötig

Mindestgröße des Aquariums

	Mindestgröße des Aquariums		Inhalt:
SS	sehr klein	20 - 40 cm	5 - 20 l
S	klein	40 - 80 cm	40 - 80 l
m	mittel	60 - 100 cm	80 - 200 l
L	groß	100 - 200 cm	200 - 400 l
XL	sehr groß	200 - 400 cm	400 - 3000 l
XXL	extrem groß	über 400 cm	über 3000 l (Schauaquarien)

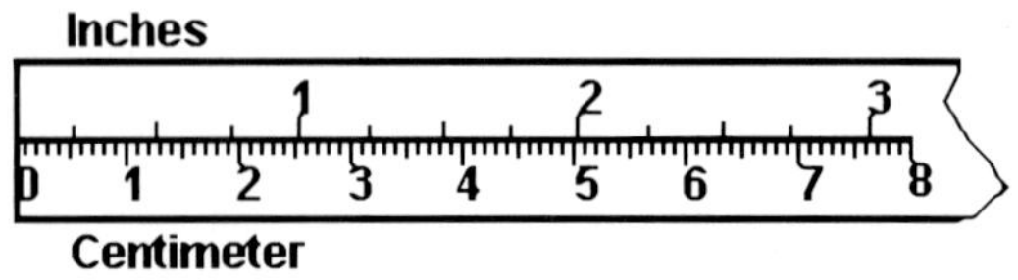

A26600-5 *Cyathochromis obliquidens* TREWAVAS, 1935

Lake Malawi: all coasts, 15 cm

Photo: W. Staeck

A26600-5 *Cyathochromis obliquidens* Trewavas, 1935

Lake Malawi: Likoma/Madimba, 15 cm

Photo: A. Spreinat

A26601-5 *Cyathochromis obliquidens* TREWAVAS, 1935

Lake Malawi: Chizumulu, 15 cm

Photo: A. Konings

A26602-5 *Cyathochromis obliquidens* TREWAVAS, 1935

Lake Malawi: Likoma, 15 cm

Photo: A. Konings

A26603-5 *Cyathochromis obliquidens* TREWAVAS, 1935

Lake Malawi: Magunga, 15 cm

Photo: A. Spreinat

A26604-5 *Cyathochromis obliquidens* TREWAVAS, 1935

Lake Malawi: Manda (Tanzania), 15 cm

Photo: A. Spreinat

A26605-5 *Cyathochromis obliquidens* TREWAVAS, 1935

Lake Malawi: Mbamba Bay, 15 cm

Photo: A. Spreinat

A26606-5 *Cyathochromis obliquidens* TREWAVAS, 1935

Lake Malawi: Otter Point, 15 cm

Photo: A. Konings

A27929-5 *Cynotilapia afra* (GÜNTHER, 1894)

Lake Malawi: Nkhata Bay, 9 cm

Photo: A.Spreinat

A27905-5 *Cynotilapia afra* (GÜNTHER, 1894)

Lake Malawi: Chitande Isl./Chilumba, 9 cm

Photo: Ad Konings

A27939-5 *Cynotilapia afra* (GÜNTHER, 1894)

Lake Malawi: Likoma Isl.: Msekwa Point, 9 cm

Photo: A.Spreinat

A27939-5 *Cynotilapia afra* (GÜNTHER, 1894)
Mutation

Lake Malawi: Likoma Isl.: Msekwa Point, 9 cm

Photo: A.Spreinat

A27925-5 *Cynotilapia afra* (GÜNTHER, 1894)

Lake Malawi: Likoma Isl.: Mbamba Isl., 9 cm

Photo: W. Deproost

A27926-5 *Cynotilapia afra* (GÜNTHER, 1894)

Lake Malawi: Metangula (Mozambique), 9 cm

Photo: Ad Konings

A27935-5 *Cynotilapia afra* (GÜNTHER, 1894)

Lake Malawi: Undu Point (Tanzania), 9 cm

Photo: A.Spreinat

A27944-5 *Cynotilapia afra* (GÜNTHER, 1894)

Lake Malawi: Ponta Messuli (Mozambique), 9 cm

Photo: A.Spreinat

A27910-5 *Cynotilapia afra* (GÜNTHER, 1894)

Lake Malawi: Mbamba Bay: Higga Reef, 9 cm

Photo: A.Spreinat

A27912-5 *Cynotilapia afra* (GÜNTHER, 1894)

Lake Malawi: Ikombe (Tanzania), 9 cm

Photo: Ad Konings

A27943-5 *Cynotilapia afra* (GÜNTHER, 1894)

Lake Malawi: Ndumbi (Tanzania), 9 cm

Photo: A.Spreinat

A27932-5 *Cynotilapia afra* (GÜNTHER, 1894)

Lake Malawi: Pombo Rocks (Tanzania), 9 cm

Photo: Ad Konings

A27938-5 *Cynotilapia afra* (GÜNTHER, 1894)

Lake Malawi: Cobue (Mozambique): Mara Point, 9 cm

Photo: A.Spreinat

A27933-5 *Cynotilapia afra* (GÜNTHER, 1894)

Lake Malawi: Puulu (Tanzania), 9 cm

Photo: Ad Konings

A27933-5 *Cynotilapia afra* (GÜNTHER, 1894)

Lake Malawi: Puulu (N) (Tanzania), 9 cm

Photo: A.Spreinat

A27922-5 *Cynotilapia afra* (GÜNTHER, 1894)

Lake Malawi: Magunga (Tanzania), 9 cm

Photo: A.Spreinat

A27908-5 *Cynotilapia afra* (GÜNTHER, 1894)

Lake Malawi: Gome Rock (at Masinje), 9 cm

Photo: Ad Konings

A27931-5 *Cynotilapia afra* (GÜNTHER, 1894)

Lake Malawi: Nkhungu Reef (Mozambique), 9 cm

Photo: E.Schraml

A27909-5 *Cynotilapia afra* (GÜNTHER, 1894)

Lake Malawi: Hai Reef (Tanzania), 9 cm

Photo: A.Spreinat

A27909-5 *Cynotilapia afra* (GÜNTHER, 1894)

Lake Malawi: Hai Reef (Tanzania), 9 cm

Photo: A.Spreinat

A27903-5 *Cynotilapia afra* (GÜNTHER, 1894)

Lake Malawi: Chewere/Chilumba, 9 cm

Photo: A.Spreinat

A27904-5 *Cynotilapia afra* (GÜNTHER, 1894)

Lake Malawi: Chisumulu, Chiwe Rocks, 9 cm

Photo: A.Spreinat

A27911-5 *Cynotilapia afra* (GÜNTHER, 1894)

Lake Malawi: Hongi Isl. (Tanzania), 9 cm

Photo: A.Spreinat

A27906-5 *Cynotilapia afra* (GÜNTHER, 1894)

Lake Malawi: Chiwindi (Mozambique), 9 cm

Photo: Ad Konings

A27915-5 *Cynotilapia afra* (GÜNTHER, 1894)

Lake Malawi: Likoma Isl.: Maingano, 9 cm
♂ Photo: W. Deproost

A27928-5 *Cynotilapia afra* (GÜNTHER, 1894)

Lake Malawi: Likoma Isl.: Ndumbi Rocks, 9 cm
♂ Photo: A.Spreinat

A27938-5 *Cynotilapia afra* (GÜNTHER, 1894)

Lake Malawi: Mara Point (Mozambique), 9 cm
♂ Photo: W.Deproost

A27936-5 *Cynotilapia afra* (GÜNTHER, 1894)

Lake Malawi: Cobue (Mozambique): Chicane, 9 cm
♂ Photo: A.Spreinat

A27937-5 *Cynotilapia afra* (GÜNTHER, 1894)

Lake Malawi: Cobue (Mozambique): Namisi Rock, 9 cm
♂ Photo: A.Spreinat

A27937-5 *Cynotilapia afra* (GÜNTHER, 1894)

Lake Malawi: Cobue (Mozambique): Namisi Rock, 9 cm
♀ Photo: A.Spreinat

A27920-5 *Cynotilapia afra* (GÜNTHER, 1894)

Lake Malawi: Lundo Isl. (Tanzania), 9 cm
♂ Photo: A.Spreinat

A27918-5 *Cynotilapia afra* (GÜNTHER, 1894)

Lake Malawi: Mbamba Bay (Tanzania): Luhuchi Rocks, 9 cm
♂ Photo: A.Spreinat

A27929-5 *Cynotilapia afra* (GÜNTHER, 1894)

Lake Malawi: Nkhata Bay, 9 cm

♂ Photo: Ad Konings

A27929-4 *Cynotilapia afra* (GÜNTHER, 1894)

Lake Malawi: Nkhata Bay, 9 cm

♀ Photo: Ad Konings

A27931-4 *Cynotilapia afra* (GÜNTHER, 1894)

Lake Malawi: Nkhungu Reef (Mozambique), 9 cm

♂ Photo: Ad Konings

A27907-5 *Cynotilapia afra* (GÜNTHER, 1894)

Lake Malawi: Cobue (Mozambique), N>Kanjindo, 9 cm

♂ Photo: A. Spreinat

A27907-5 *Cynotilapia afra* (GÜNTHER, 1894)

Lake Malawi: Cobue (Mozambique), 9 cm

♂ Photo: Ad Konings

A27907-4 *Cynotilapia afra* (GÜNTHER, 1894)

Lake Malawi: Cobue (Mozambique), 9 cm

♀ Photo: Ad Konings

A27927-5 *Cynotilapia afra* (GÜNTHER, 1894)

Lake Malawi: Minos Reef (Mozambique), 9 cm

♂ Photo: Ad Konings

A27927-4 *Cynotilapia afra* (GÜNTHER, 1894)

Lake Malawi: Minos Reef (Mozambique), 9 cm

♀ Photo: Ad Konings

A27946-5 *Cynotilapia afra* (GÜNTHER, 1894)

Lake Malawi: Tumbi Point (Tanzania), 9 cm

Photo: Ad Konings

A28902-5 *Cynotilapia afra* (GÜNTHER, 1894)

Lake Malawi: Cape Kaiser, Tanzania, 9 cm

Photo: Ad Konings

A27934-5 *Cynotilapia afra* (GÜNTHER, 1894)

Lake Malawi: Thumbi West Isl., 9 cm

Photo: A.Spreinat

A27914-5 *Cynotilapia afra* (GÜNTHER, 1894)

Lake Malawi: Kakusa, 9 cm

Photo: Ad Konings

A27942-5 *Cynotilapia afra* (GÜNTHER, 1894)

Lake Malawi: Njambe (Tanzania)

Photo: A.Spreinat

A27917-5 *Cynotilapia afra* (GÜNTHER, 1894)

Lake Malawi: Londo (Mozambique), 9 cm

Photo: Ad Konings

A27916-5 *Cynotilapia afra* (GÜNTHER, 1894)

Lake Malawi: Lion's Cove, 9 cm

Photo: A.Spreinat

A27924-5 *Cynotilapia afra* (GÜNTHER, 1894)

Lake Malawi: Mbahwa Isl. (Tanzania), 9 cm

Photo: A.Spreinat

A27940-5 *Cynotilapia afra* (GÜNTHER, 1894)

Lake Malawi: Likoma Isl.: Mazimbwe Isl., 9 cm

Photo: A.Spreinat

A27941-5 *Cynotilapia afra* (GÜNTHER, 1894)

Lake Malawi: Likoma Isl.: Mbuzi Isl., 9 cm

Photo: A.Spreinat

A27923-5 *Cynotilapia afra* (GÜNTHER, 1894)

Lake Malawi: Mbamba Bay (Tanzania): Mara Rocks, 9 cm

Photo: A.Spreinat

A28025-4 *Cynotilapia* spec. "Thomasi"
(also often called *Cynotilapia* spec. "Mbweca")
Lake Malawi: 8-10 cm

Photo: W.Staeck

A28025-5 *Cynotilapia* spec. "Thomasi"
(also often called *Cynotilapia* spec. "Mbweca")
Lake Malawi: Mbweca (Mozambique), 8-10 cm

Photo: Ad Konings

A28025-5 *Cynotilapia* spec. "Thomasi"
(also often called *Cynotilapia* spec. "Mbweca")
Lake Malawi: Cobue: S-Mara Point (Mozambique), 8-10 cm

Photo: A.Spreinat

A27970-5 *Cynotilapia* spec. "Black Dorsal"

Lake Malawi: Mbenji Isl., 10 cm

Photo: Ad Konings

A27970-5 *Cynotilapia* spec. "Black Dorsal"

Lake Malawi: Mbenji Isl., 10 cm

Photo: Ad Konings

A27980-5 *Cynotilapia* spec. "Black Eastern"

Lake Malawi: Cobue, Kanjindo (Mozambique), 9 cm

Photo: A.Spreinat

A27980-5 *Cynotilapia* spec. "Black Eastern"

Lake Malawi: Cobue, Kanjindo (Mozambique), 9 cm

Photo: A.Spreinat

A27982-5 *Cynotilapia* spec. "Black Eastern"

Lake Malawi: Makanjila, 9 cm

Photo: A.Spreinat

A27981-5 *Cynotilapia* spec. "Black Eastern"

Lake Malawi: Gome, 9 cm

Photo: Ad Konings

A27984-5 *Cynotilapia* spec. "Black Eastern"

Lake Malawi: Nkhungu Point (Mozambique), 9 cm

Photo: Ad Konings

A27984-5 *Cynotilapia* spec. "Black Eastern"

Lake Malawi: Nkhungu Point (Mozambique), 9 cm

Photo: Ad Konings

A27985-5 *Cynotilapia* spec. "Chinyankwazi"

Lake Malawi: Chinyankwazi

Photo: A. Spreinat

A28023-5 *Cynotilapia* spec. "Mbamba"

Lake Malawi: Chisumulu: Chilanje, 8-10 cm

Photo: A.Spreinat

A27997-5 *Cynotilapia* spec. "Lion"

Lake Malawi: Lion's Cove, 8-10 cm

Photo: E.Schraml

A27997-4 *Cynotilapia* spec. "Lion"

Lake Malawi: Lion's Cove, 8-10 cm

Photo: E.Schraml

A27998-5 *Cynotilapia* spec. "Lion"

Lake Malawi: Lundu (Tanzania), 8-10 cm

Photo: A.Spreinat

A27999-5 *Cynotilapia* spec. "Lion"

Lake Malawi: Lupingu (Tanzania), 8-10 cm

Photo: Ad Konings

A28000-5 *Cynotilapia* spec. "Lion"

Lake Malawi: Magunga (Tanzania), 8-10 cm

Photo: A.Spreinat

A28001-5 *Cynotilapia* spec. "Lion"

Lake Malawi: Manda (Tanzania), 8-10 cm

Photo: A.Spreinat

A28010-5 *Cynotilapia* spec. "Maleri"

Lake Malawi: Maleri Isl., 8-10 cm

Photo: Ad Konings

A28010-5 *Cynotilapia* spec. "Maleri"

Lake Malawi: Maleri Isl., 8-10 cm

Photo: Ad Konings

A28016-5 *Cynotilapia* spec. "Mbamba"

Lake Malawi: Chitendi Isl., 8-10 cm

Photo: A.Spreinat

A28017-5 *Cynotilapia* spec. "Mbamba"

Lake Malawi: Kakusa, 8-10 cm

Photo: Ad Konings

A28022-5 *Cynotilapia* spec. "Mbamba"

Lake Malawi: Lion's Cove (Tanzania), 8-10 cm

Photo: Ad Konings

A28018-5 *Cynotilapia* spec. "Mbamba"

Lake Malawi: Lundu (Tanzania), 8-10 cm

Photo: Ad Konings

A28020-5 *Cynotilapia* spec. "Mbamba"

Lake Malawi: Mphanga Rocks, 8-10 cm

Photo: Ad Konings

A28021-5 *Cynotilapia* spec. "Mbamba"

Lake Malawi: Nkhata Bay, 8-10 cm

Photo: A.Spreinat

A28021-5 *Cynotilapia* spec. "Mbamba"

Lake Malawi: Nkhata Bay, 8-10 cm

Photo: Ad Konings

A28021-3 *Cynotilapia* spec. "Mbamba"

Lake Malawi: Nkhata Bay, 8-10 cm

Photo: Ad Konings

A28031-5 *Cynotilapia* spec. "Ndumbi"

Lake Malawi: Chilucha Reef (Mozambique), 8-10 cm

Photo: Ad Konings

A28032-5 *Cynotilapia* spec. "Ndumbi"

Lake Malawi: Lupingu (Tanzania), 8-10 cm

Photo: Ad Konings

A27951-5 *Cynotilapia axelrodi* Burgess, 1976

Lake Malawi: Nkhata Bay, 9 cm

Photo: A.Spreinat

A27951-5 *Cynotilapia axelrodi* Burgess, 1976

Lake Malawi: Nkhata Bay, 9 cm

Photo: A.Spreinat

A28036-5 *Cynotilapia* spec. „Jalo" leg. Spreinat
= *Cynotilapia* spec. „Yellow Dorsal" leg. Konings
Lake Malawi: Jalo/Jaro Reef, 9 cm

Photo: A.Spreinat

A28037-5 *Cynotilapia* spec. "Yellow Dorsal" leg. Konings
= *Cynotilapia* spec. „Jalo" leg. Spreinat
Lake Malawi: Mbenji Isl., 8-10 cm

Photo: Ad Konings

A28036-5 *Cynotilapia* spec. "Yellow Dorsal" leg. Konings
= *Cynotilapia* spec. „Jalo" leg. Spreinat
Lake Malawi: Jalo Reef, 8-10 cm

Photo: Ad Konings

A28036-5 *Cynotilapia* spec. "Yellow Dorsal" leg. Konings
= *Cynotilapia* spec. „Jalo" leg. Spreinat
Lake Malawi: Jalo Reef, 8-10 cm

Photo: Ad Konings

A32356-5 *Genyochromis mento* TREWAVAS, 1935
Malawisee-Schuppenfresser/ Malawian Scale Eater
Lake Malawi: Mbamba Bay (Tanzania), 12 cm
Photo: L.Seegers

A32350-5 *Genyochromis mento* TREWAVAS, 1935
Malawisee-Schuppenfresser/ Malawian Scale Eater
Lake Malawi, 12 cm
Photo: W.Staeck

A32351-4 *Genyochromis mento* TREWAVAS, 1935
Malawisee-Schuppenfresser/ Malawian Scale Eater
Lake Malawi: Chadagha, 12 cm
Photo: Ad Konings

A32352-5 *Genyochromis mento* TREWAVAS, 1935
Malawisee-Schuppenfresser/ Malawian Scale Eater
Lake Malawi: Chinyankwazi, 12 cm
Photo: Ad Konings

A32353-5 *Genyochromis mento* TREWAVAS, 1935
Malawisee-Schuppenfresser/ Malawian Scale Eater
Lake Malawi: Kande Isl., 12 cm
Photo: Ad Konings

A32360-5 *Genyochromis mento* TREWAVAS, 1935
Malawisee-Schuppenfresser/ Malawian Scale Eater, OB
Lake Malawi: Cobue (Mozambique): Mara Point, 12 cm
Photo: A.Spreinat

A32354-5 *Genyochromis mento* TREWAVAS, 1935
Malawisee-Schuppenfresser/ Malawian Scale Eater
Lake Malawi: Likoma Isl., 12 cm
Photo: Ad Konings

A32355-5 *Genyochromis mento* TREWAVAS, 1935
Malawisee-Schuppenfresser/ Malawian Scale Eater
Lake Malawi: Matema (Tanzania), 12 cm
Photo: L.Seegers

A32470-5 *Gephyrochromis* spec. "Zebroides"

Lake Malawi: Lingwezi (Mozambique), 12 cm ♂

Photo: A.Spreinat

A32470-4 *Gephyrochromis* spec. "Zebroides"

Lake Malawi: Lumessi (Mozambique), 12 cm

Photo: Ad Konings

A38415-5 *Iodotropheus* spec. "Matema"

Lake Malawi: Matema (Tanzania), 10 cm ♂

Photo: L.Seegers

A38415-5 *Iodotropheus* spec. "Matema"

Lake Malawi: Lumbira (Tanzania), 10 cm ♀

Photo: L.Seegers

A38420-5 *Iodotropheus sprengerae* Oliver & Loiselle, 1972

Lake Malawi: Aquariumspecimen, 10 cm ♂

Photo: W.Deproost

A38420-5 *Iodotropheus sprengerae* Oliver & Loiselle, 1972

Lake Malawi: Aquariumspecimen, 10 cm ♀

Photo: Ch.Wyrwich

A38421-5 *Iodotropheus sprengerae* Oliver & Loiselle, 1972

Lake Malawi: Chinyamwezi, 10 cm ♂

Photo: Ad Konings

A38420-5 *Iodotropheus sprengerae* Oliver & Loiselle, 1972

Lake Malawi: Aquariumspecimen, 10 cm ♂

Photo: A.Spreinat

A38420-5 *Iodotropheus sprengerae* Oliver & Loiselle, 1972

Lake Malawi: Aquarium specimen, 10 cm

photo: E. Schraml

A38430-4 *Iodotropheus declivitas* Stauffer, 1994

Lake Malawi: Boadzulu Isl., 10 cm

photo: J. R. Stauffer
from: Stauffer, 1994. With kind permission of the author and the publisher (Verlag Dr. F. Pfeil)

A38425-5 *Iodotropheus stuartgranti* Konings, 1990

Lake Malawi: Gome at Nsinje River, 8-10 cm

Photo: Ad Konings

A38425-5 *Iodotropheus stuartgranti* Konings, 1990

Lake Malawi: Gome at Nsinje River, 8-10 cm

Photo: Ad Konings

Labeotropheus fuelleborni im natürlichen Lebenraum: Cove Mountain, Manda, Tanzania

Labeotropheus fuelleborni in its natural habitat: Cove Mountain, Manda, Tanzania

Photo: A. Spreinat

A39801-5 *Labeotropheus fuelleborni* AHL, 1926
Schabemaulbuntbarsch / Fuelleborn´s Cichlid
Lake Malawi: Chidunga Isl., 15 cm
Photo: Ad Konings

A39815-5 *Labeotropheus fuelleborni* AHL, 1926
Schabemaulbuntbarsch / Fuelleborn´s Cichlid
Lake Malawi: Mbenji Isl., 15 cm
Photo: W.Staeck

A39828-5 *Labeotropheus fuelleborni* AHL, 1926
Schabemaulbuntbarsch / Fuelleborn´s Cichlid
Lake Malawi: Chilumba: Chirwa Isl., 15 cm
Photo: A.Spreinat

A39809-5 *Labeotropheus fuelleborni* AHL, 1926
Schabemaulbuntbarsch / Fuelleborn´s Cichlid
Lake Malawi: Chilumba: Katale Isl., 15 cm
Photo: Ad Konings

A39807-5 *Labeotropheus fuelleborni* AHL, 1926
Schabemaulbuntbarsch / Fuelleborn´s Cichlid
Lake Malawi: Hora Mhango, 15 cm
Photo: Ad Konings

A39823-5 *Labeotropheus fuelleborni* AHL, 1926
Schabemaulbuntbarsch / Fuelleborn´s Cichlid
Lake Malawi: Selewa, 15 cm
Photo: Ad Konings

A39816-5 *Labeotropheus fuelleborni* AHL, 1926
Schabemaulbuntbarsch / Fuelleborn´s Cichlid
Lake Malawi: Mdoka, 15 cm
Photo: Ad Konings

A39819-5 *Labeotropheus fuelleborni* AHL, 1926
Schabemaulbuntbarsch / Fuelleborn´s Cichlid
Lake Malawi: Nakantenga, 15 cm
Photo: Ad Konings

A39810-5 *Labeotropheus fuelleborni* X *Maylandia zebra*
Kreuzung mit *Maylandia zebra* / Cross-breed
kein natürliches Vorkommen/ does not appear in nature
Photo: E.Schraml

A39813-5 *Labeotropheus fuelleborni* AHL, 1926
Schabemaulbuntbarsch / Fuelleborn´s Cichlid
Lake Malawi: Manda: Cove Mountain, 15 cm
Photo: A.Spreinat

A39811-5 *Labeotropheus fuelleborni* AHL, 1926
Schabemaulbuntbarsch / Fuelleborn´s Cichlid
Lake Malawi: Likoma: Ndumbi Rocks, 15 cm
Photo: A.Spreinat

A39811-5 *Labeotropheus fuelleborni* AHL, 1926
Schabemaulbuntbarsch / Fuelleborn´s Cichlid, OB-Morph
Lake Malawi: Likoma Isl., 15 cm
Photo:

A39826-5 *Labeotropheus fuelleborni* AHL, 1926
Schabemaulbuntbarsch / Fuelleborn´s Cichlid
Lake Malawi: Tumbi Point (Mozambique), 15 cm
Photo: Ad Konings

A39825-5 *Labeotropheus fuelleborni* AHL, 1926
Schabemaulbuntbarsch / Fuelleborn´s Cichlid
Lake Malawi: Thumbi W. Isl., 15 cm
Photo: A.Spreinat

A39817-5 *Labeotropheus fuelleborni* AHL, 1926
Schabemaulbuntbarsch / Fuelleborn´s Cichlid
Lake Malawi: Metangula (Mozambique), 15 cm
Photo: Ad Konings

A39827-5 *Labeotropheus fuelleborni* AHL, 1926
Schabemaulbuntbarsch / Fuelleborn´s Cichlid
Lake Malawi: Zimbawe Rock, 15 cm
Photo: Ad Konings

A39814-5 *Labeotropheus fuelleborni* Ahl, 1926
Schabemaulbuntbarsch / Fuelleborn´s Cichlid
Lake Malawi: Mara, 15 cm
Photo: E.Schraml

A39821-5 *Labeotropheus fuelleborni* Ahl, 1926
Schabemaulbuntbarsch / Fuelleborn´s Cichlid
Lake Malawi: Nkhata Bay, 15 cm
Photo: Ad Konings

A38820-5 *Labeotropheus fuelleborni* Ahl, 1926
Schabemaulbuntbarsch / Fuelleborn´s Cichlid
Lake Malawi: Nankoma, 15 cm
Photo: E.Schraml

A39822-5 *Labeotropheus fuelleborni* Ahl, 1926
Schabemaulbuntbarsch / Fuelleborn´s Cichlid, O-Morph
Lake Malawi, 15 cm
Photo: E.Schraml

A39803-6 *Labeotropheus fuelleborni* Ahl, 1926
Schabemaulbuntbarsch / Fuelleborn´s Cichlid, O-Morph
Lake Malawi: Chinyamwezi, 15 cm
Photo: W.Staeck

A39822-6 *Labeotropheus fuelleborni* Ahl, 1926
Schabemaulbuntbarsch / Fuelleborn´s Cichlid, OB-Morph
Lake Malawi, 15 cm
Photo: W.Staeck

A39840-5 *Labeotropheus spec.* "Pernostus"
=? *Labeotropheus trewavasae* (Gome)
Lake Malawi: 12 cm
Photo: E.Schraml

A39881-5 *Labeotropheus trewavasae* Fryer, 1956
Gestreckter Schabemaulbuntbarsch / Red Top Cichlid
Lake Malawi: Gome, 12 cm
Photo: Ad Konings

A39853-5 *Labeotropheus trewavasae* Fryer, 1956
Gestreckter Schabemaulbuntbarsch / Red Top Cichlid
Lake Malawi: Charo, 12 cm
♂ Photo: Ad Konings

A39862-5 *Labeotropheus trewavasae* Fryer, 1956
Gestreckter Schabemaulbuntbarsch / Red Top Cichlid
Lake Malawi: Lundu (Tanzania), 12 cm
♂ Photo: Ad Konings

A39866-5 *Labeotropheus trewavasae* Fryer, 1956
Gestreckter Schabemaulbuntbarsch / Red Top Cichlid
Lake Malawi: Masinje, 12 cm
♂ Photo: Ad Konings

A39866-5 *Labeotropheus trewavasae* Fryer, 1956
Gestreckter Schabemaulbuntbarsch / Red Top Cichlid, OB-Morph
Lake Malawi: Masinje, 12 cm
♀ Photo: Ad Konings

A39855-5 *Labeotropheus trewavasae* Fryer, 1956
Gestreckter Schabemaulbuntbarsch / Red Top Cichlid
Lake Malawi: Chitande, 12 cm
♂ Photo: W.Staeck

A39870-5 *Labeotropheus trewavasae* Fryer, 1956
Gestreckter Schabemaulbuntbarsch / Red Top Cichlid
Lake Malawi: Mozambique, 12 cm
♂ Photo: W.Staeck

A39854-5 *Labeotropheus trewavasae* Fryer, 1956
Gestreckter Schabemaulbuntbarsch / Red Top Cichlid
Lake Malawi: Chilumba, 12 cm
♂ Photo: E.Schraml

A39854-5 *Labeotropheus trewavasae* Fryer, 1956
Gestreckter Schabemaulbuntbarsch / Red Top Cichlid
Lake Malawi: Chilumba, 12 cm
♂ Photo: W.Staeck

A39854-5 *Labeotropheus trewavasae* FRYER, 1956
Gestreckter Schabemaulbuntbarsch / Red Top Cichlid
Lake Malawi: Chilumba, 12 cm
Photo: E.Schraml

A39854-5 *Labeotropheus trewavasae* FRYER, 1956
Gestreckter Schabemaulbuntbarsch / Red Top Cichlid
Lake Malawi: Chilumba, 12 cm
Photo: A.Spreinat

A39886-5 *Labeotropheus trewavasae* FRYER, 1956
Gestreckter Schabemaulbuntbarsch / Red Top Cichlid
Lake Malawi: Cobue: Mara Point (Mozambique), 12 cm
Photo: A.Spreinat

A39887-5 *Labeotropheus trewavasae* FRYER, 1956
Gestreckter Schabemaulbuntbarsch / Red Top Cichlid
Lake Malawi: Ponta Messuli (Mozambique), 12 cm
Photo: A.Spreinat

A39888-5 *Labeotropheus trewavasae* FRYER, 1956
Gestreckter Schabemaulbuntbarsch / Red Top Cichlid
Lake Malawi: Tumbi Point (Mozambique), 12 cm
Photo: Ad Konings

A39868-5 *Labeotropheus trewavasae* FRYER, 1956
Gestreckter Schabemaulbuntbarsch / Red Top Cichlid
Lake Malawi: Metangula (Mozambique), 12 cm
Photo: Ad Konings

A39877-5 *Labeotropheus trewavasae* FRYER, 1956
Gestreckter Schabemaulbuntbarsch / Red Top Cichlid
Lake Malawi: Puulu (Tanzania), 12 cm
Photo: A.Spreinat

A39867-5 *Labeotropheus trewavasae* FRYER, 1956
Gestreckter Schabemaulbuntbarsch / Red Top Cichlid
Lake Malawi: Mdoka, 12 cm
Photo: Ad Konings

A39859-5 *Labeotropheus trewavasae* FRYER, 1956
Gestreckter Schabemaulbuntbarsch / Red Top Cichlid
Lake Malawi: Lion's Cove, 12 cm
Photo: Ad Konings

A39876-5 *Labeotropheus trewavasae* FRYER, 1956
Gestreckter Schabemaulbuntbarsch / Red Top Cichlid
Lake Malawi: Nkhata Bay, 12 cm
Photo: E.Schraml

A39858-5 *Labeotropheus trewavasae* FRYER, 1956
Gestreckter Schabemaulbuntbarsch / Red Top Cichlid
Lake Malawi: Kirondo (Tanzania), 12 cm
Photo: A.Spreinat

A39873-5 *Labeotropheus trewavasae* FRYER, 1956
Gestreckter Schabemaulbuntbarsch / Red Top Cichlid
Lake Malawi: Nkhata Bay, 12 cm
Photo: Ad Konings

A39871-5 *Labeotropheus trewavasae* FRYER, 1956
Gestreckter Schabemaulbuntbarsch / Red Top Cichlid
Lake Malawi: Mphanga Rocks, 12 cm
Photo: Ad Konings

A39885-5 *Labeotropheus trewavasae* FRYER, 1956
Gestreckter Schabemaulbuntbarsch / Red Top Cichlid
Lake Malawi: Mumbo Isl., 12 cm
Photo: A.Spreinat

A39889-5 *Labeotropheus trewavasae* FRYER, 1956
Gestreckter Schabemaulbuntbarsch / Red Top Cichlid
Lake Malawi: Nakanthenga, 12 cm
Photo: A.Spreinat

A39889-5 *Labeotropheus trewavasae* FRYER, 1956
Gestreckter Schabemaulbuntbarsch / Red Top Cichlid, OB-Morph
Lake Malawi: Nakanthenga, 12 cm
Photo: A.Spreinat

A39882-5 *Labeotropheus trewavasae* FRYER, 1956
Gestreckter Schabemaulbuntbarsch / Red Top Cichlid
Lake Malawi: Lupingu, 12 cm
Photo: Ad Konings

A39861-5 *Labeotropheus trewavasae* FRYER, 1956
Gestreckter Schabemaulbuntbarsch / Red Top Cichlid
Lake Malawi: Lumessi (Mozambique), 12 cm
Photo: Ad Konings

A39869-5 *Labeotropheus trewavasae* FRYER, 1956
Gestreckter Schabemaulbuntbarsch / Red Top Cichlid, OB-Moph
Lake Malawi: Minos Reef (Mozambique), 12 cm
Photo: Ad Konings

A39869-5 *Labeotropheus trewavasae* FRYER, 1956
Gestreckter Schabemaulbuntbarsch / Red Top Cichlid
Lake Malawi: Minos Reef (Mozambique), 12 cm
Photo: Ad Konings

A39878-5 *Labeotropheus trewavasae* FRYER, 1956
Gestreckter Schabemaulbuntbarsch / Red Top Cichlid
Lake Malawi: Taiwan Reef , 12 cm
Photo: Ad Konings

A39878-5 *Labeotropheus trewavasae* FRYER, 1956
Gestreckter Schabemaulbuntbarsch / Red Top Cichlid, OB-Morph
Lake Malawi: Taiwan Reef , 12 cm
Photo: Ad Konings

A39884-5 *Labeotropheus trewavasae* FRYER, 1956
Gestreckter Schabemaulbuntbarsch / Red Top Cichlid, OB-Morph
Lake Malawi: Pombo Rocks (Tanzania), 12 cm
Photo: Ad Konings

A39856-5 *Labeotropheus trewavasae* FRYER, 1956
Gestreckter Schabemaulbuntbarsch / Red Top Cichlid, OB-Morph
Lake Malawi: Mbamba Bay (Tanzania), Higga Reef , 12 cm
Photo: A.Spreinat

A39856-5 *Labeotropheus trewavasae* FRYER, 1956
Gestreckter Schabemaulbuntbarsch / Red Top Cichlid
Lake Malawi: Mbamba Bay (Tanzania), Higga Reef , 12 cm
♂ Photo: Ad Konings

A39856-5 *Labeotropheus trewavasae* FRYER, 1956
Gestreckter Schabemaulbuntbarsch / Red Top Cichlid, OB-Morph
Lake Malawi: Mbamba Bay (Tanzania), Higga Reef , 12 cm
♀ Photo: Ad Konings

A39872-5 *Labeotropheus trewavasae* FRYER, 1956
Gestreckter Schabemaulbuntbarsch / Red Top Cichlid
Lake Malawi: Ngkuyo Isl. (Tanzania), 12 cm
♂ Photo: A.Spreinat

A39860-5 *Labeotropheus trewavasae* FRYER, 1956
Gestreckter Schabemaulbuntbarsch / Red Top Cichlid
Lake Malawi: Lumbira (Tanzania), 12 cm
♂ Photo: L.Seegers

A39865-5 *Labeotropheus trewavasae* FRYER, 1956
Gestreckter Schabemaulbuntbarsch / Red Top Cichlid
Lake Malawi: Mbamba Bay (Tanzania), Mara Rocks , 12 cm
♂ Photo: A.Spreinat

A39863-5 *Labeotropheus trewavasae* FRYER, 1956
Gestreckter Schabemaulbuntbarsch / Red Top Cichlid
Lake Malawi: Maleri Isl. , 12 cm
♂ Photo: W.Staeck

A39879-5 *Labeotropheus trewavasae* FRYER, 1956
Gestreckter Schabemaulbuntbarsch / Red Top Cichlid, OB-Morph
Lake Malawi: Thumbi W. Isl. , 12 cm
♂ Photo: E.Schraml

A39879-5 *Labeotropheus trewavasae* FRYER, 1956
Gestreckter Schabemaulbuntbarsch / Red Top Cichlid, OB-Morph
Lake Malawi: Thumbi W. Isl. , 12 cm
♀ Photo: A.Spreinat

A39879-5 *Labeotropheus trewavasae* FRYER, 1956
Gestreckter Schabemaulbuntbarsch / Red Top Cichlid, OB-Morph
Lake Malawi: Thumbi W. Isl. , 12 cm
♂ Photo: E.Schraml

A39879-5 *Labeotropheus trewavasae* FRYER, 1956
Gestreckter Schabemaulbuntbarsch / Red Top Cichlid
Lake Malawi: Thumbi W. Isl. , 12 cm
♂ Photo: E.Schraml

A39879-5 *Labeotropheus trewavasae* FRYER, 1956
Gestreckter Schabemaulbuntbarsch / Red Top Cichlid
Lake Malawi: Thumbi W. Isl. , 12 cm
♂ Photo: A.Spreinat

A39879-5 *Labeotropheus trewavasae* FRYER, 1956
Gestreckter Schabemaulbuntbarsch / Red Top Cichlid
Lake Malawi: Thumbi W. Isl. , 12 cm
♀ Photo: A.Spreinat

A39879-5 *Labeotropheus trewavasae* FRYER, 1956
Gestreckter Schabemaulbuntbarsch / Red Top Cichlid, O-Morph
Lake Malawi: Thumbi W. Isl. , 12 cm
♂ Photo: E.Schraml

A39879-5 *Labeotropheus trewavasae* FRYER, 1956
Gestreckter Schabemaulbuntbarsch / Red Top Cichlid, O-Morph
Lake Malawi: Thumbi W. Isl. , 12 cm
♀ Photo: W.Staeck

A39851-5 *Labeotropheus trewavasae* FRYER, 1956
Gestreckter Schabemaulbuntbarsch / Red Top Cichlid
Lake Malawi: Boadzulu , 12 cm
♂ Photo: Ad Konings

A39883-5 *Labeotropheus trewavasae* FRYER, 1956
Gestreckter Schabemaulbuntbarsch / Red Top Cichlid
Lake Malawi: Namalenje , 12 cm
♂ Photo: Ad Konings

A39875-5 *Labeotropheus trewavasae* FRYER, 1956
Gestreckter Schabemaulbuntbarsch / Red Top Cichlid, OB-Form
Lake Malawi: Tanzania , 12 cm
Photo: M.K.Meyer

A39880-5 *Labeotropheus trewavasae* FRYER, 1956
Gestreckter Schabemaulbuntbarsch / Red Top Cichlid
Lake Malawi: Chimoto Rock (Tanzania) , 12 cm
Photo: M.K.Meyer

A39852-5 *Labeotropheus trewavasae* FRYER, 1956
Gestreckter Schabemaulbuntbarsch / Red Top Cichlid
Lake Malawi: unknown locality , 12 cm
Photo: Nakano/A.C.S.

A39852-5 *Labeotropheus trewavasae* FRYER, 1956
Gestreckter Schabemaulbuntbarsch / Red Top Cichlid
Lake Malawi: unknown locality , 12 cm
Photo: Nakano/A.C.S.

A40020-5 *Labidochromis* spec. "Likomae"

Lake Malawi: Likoma Isl., 8 cm
Photo: Ad Konings

A40040-5 *Labidochromis* spec. "Luhuchi"

Lake Malawi: Mbamba Bay: Luhuchi Rocks (Tanzania), 8-9 cm
Photo: A.Spreinat

A39950-5 *Labidochromis* spec. "Deep Body"

Lake Malawi: Pombo Reef (Tanzania), 9-10 cm
Photo: A.Spreinat

A39965-5 *Labidochromis gigas* LEWIS, 1982

Lake Malawi: Likoma: Ndumbi Rocks,8-9 cm
Photo: A.Spreinat

A39965-5 *Labidochromis gigas* Lewis, 1982

Lake Malawi: Chisumulu: Chilanje, 8-9 cm

Photo: A.Spreinat

A39965-5 *Labidochromis gigas* Lewis, 1982

Lake Malawi: Likoma: Maingano, 8-9 cm

Photo: A.Spreinat

A39980-5 *Labidochromis* spec. "Gigas Cobue"

Lake Malawi: Mbweca (Mozambique), 9 cm

Photo: Ad Konings

A39985-5 *Labidochromis* spec. "Gigas Mara"

Lake Malawi: Mara Rocks, 9 cm

Photo: Ad Konings

A39970-5 *Labidochromis* spec. "Gigas Chidunga"

Lake Malawi: Chidunga Rocks, 9 cm

Photo: Ad Konings

A39970-5 *Labidochromis* spec. "Gigas Chidunga"

Lake Malawi: Chidunga Rocks, 9 cm

Photo: Ad Konings

A39975-5 *Labidochromis* spec. "Gigas Chilumba"

Lake Malawi: Katale Isl., 9 cm

Photo: Ad Konings

A39975-5 *Labidochromis* spec. "Gigas Chilumba"

Lake Malawi: Katale Isl., 9 cm

Photo: Ad Konings

A40030-5 *Labidochromis* spec. "Lividus Mozambique"

Lake Malawi: Cubue (Mozambique), 9 cm

Photo: Ad Konings

A40030-4 *Labidochromis* spec. "Lividus Mozambique"

Lake Malawi: Cobue (Mozambique), 9 cm

Photo: Ad Konings

A40035-5 *Labidochromis* spec. "Lividus Nkhungu"

Lake Malawi: Nkhungu Point (Mozambique), 9 cm

Photo: Ad Konings

A40035-5 *Labidochromis* spec. "Lividus Nkhungu"

Lake Malawi: Nkhungu Point (Mozambique), 9 cm

Photo: Ad Konings

A39990-5 *Labidochromis heterodon* Lewis, 1982

Lake Malawi: Boadzulu Isl.,9 cm

Photo: A.Spreinat

A39990-5 *Labidochromis heterodon* Lewis, 1982

Lake Malawi: Boadzulu Isl., 9 cm

Photo: A.Spreinat

A39990-5 *Labidochromis heterodon* Lewis, 1982

Lake Malawi: Boadzulu Isl., 9 cm

Photo: Ad Konings

A40065-5 *Labidochromis* spec. "Matema"

Lake Malawi: Matema (Tanzania), 9 cm

Photo: L.Seegers

A40085-5 *Labidochromis mbenjii* LEWIS, 1982

Lake Malalwi: Mbenji Isl., 9 cm

♂ Photo: A.Spreinat

A40085-5 *Labidochromis mbenjii* LEWIS, 1982

Lake Malawi: Mbenji Isl., 9 cm

♀ Photo: A.Spreinat

A40130-5 *Labidochromis vellicans* TREWAVAS, 1935

Lake Malawi: S-Lake, 8-9 cm

♂ Photo: A.Spreinat

A40130-5 *Labidochromis vellicans* TREWAVAS, 1935

Lake Malawi: S-Lake, 8-9 cm

♀ Photo: A.Spreinat

A40105-5 *Labidochromis shiranus* LEWIS, 1982

Lake Malawi: S-Lake, Shire River, 8-9 cm

♂ Photo: Ad Konings

A40015-5 *Labidochromis ianthinus* LEWIS, 1982

Lake Malawi: Mbenji Isl., Nkhomo Reef

Photo: A.Spreinat

A40110-5 *Labidochromis strigatus* LEWIS, 1982

Lake Malawi: Chizumulu Isl.: Chiteko, 7 cm

♂ Photo: A.Spreinat

A40110-5 *Labidochromis strigatus* LEWIS, 1982

Lake Malawi: Chizumulu Isl.: Chilanje, 7 cm

♂ Photo: A.Spreinat

A40050-5 *Labidochromis maculicauda* Lewis, 1982

Lake Malawi: Chilumba: Mpanga Rocks, 9 cm

♂ Photo: A.Spreinat

A40050-5 *Labidochromis maculicauda* Lewis, 1982

Lake Malawi: Nkhata Bay, 9 cm

♂ Photo: W.Deproost

A40050-5 *Labidochromis maculicauda* Lewis, 1982

Lake Malawi: Chinula (Tanzania): Mbamba Bay, 9 cm

♂ Photo: A.Spreinat

A40050-5 *Labidochromis maculicauda* Lewis, 1982

Lake Malawi: Lundo Isl., 9 cm

♂ Photo: A.Spreinat

A39955-5 *Labidochromis flavigulis* Lewis, 1982

Lake Malawi: Chisumulu, 9 cm

♂ Photo: A.Spreinat

A39955-5 *Labidochromis flavigulis* Lewis, 1982

Lake Malawi: Chisumulu, 9 cm

♀ Photo: A.Spreinat

A40120-5 *Labidochromis* spec. "Textilis Blue"

Lake Malawi: Lumessi North (Mozambique), 9 cm

♂ Photo: Ad Konings

A40125-5 *Labidochromis* spec. "Textilis Cobalt"

Lake Malawi: Lumessi South (Mozambique), 9 cm

♂ Photo: Ad Konings

A40115-5 *Labidochromis textilis* OLIVER, 1975

Lake Malawi: Lumbaulo (Mozambique), 9 cm

♂ Photo: Ad Konings

A40115-5 *Labidochromis textilis* OLIVER, 1975

Lake Malawi: Londo (Mozambique), 9 cm

♂ Photo: Ad Konings

A40116-5 *Labidochromis* cf. *textilis* OLIVER, 1975

Lake Malawi: Metangula (Mozambique): Chuanga, 9 cm

♀ Photo: A.Spreinat

A40116-5 *Labidochromis* cf. *textilis* OLIVER, 1975

Lake Malawi: Meponda (Mozambique): Nyamizimu, 9 cm

♂ Photo: A.Spreinat

A40135-5 *Labidochromis* spec. "Zebra Eastern"

Lake Malawi: Gome, 7 cm

♂ Photo: Ad Konings

A40135-5 *Labidochromis* spec. "Zebra Eastern"

Lake Malawi: Lumessi South (Mozambique), 7 cm

♂ Photo: Ad Konings

A40135-5 *Labidochromis* spec. "Zebra Eastern"

Lake Malawi: Ponta Messuli North (Mozambique), 7 cm

♂ Photo: A.Spreinat

A40135-5 *Labidochromis* spec. "Zebra Eastern"

Lake Malawi: Chiloelo (Mozambique): Ngulu, 7 cm

♂ Photo: A.Spreinat

A40140-5 *Labidochromis zebroides* LEWIS, 1982

Lake Malawi: Likoma: Mazimbwe Isl., 7 cm

Photo: A.Spreinat

A39960-5 *Labidochromis freibergi* JOHNSON, 1974

Lake Malawi: Likoma, 8 cm

Photo: W.Deproost

A39960-5 *Labidochromis freibergi* JOHNSON, 1974

Lake Malawi: Likoma, 8 cm

Photo: A.Spreinat

A39960-5 *Labidochromis freibergi* JOHNSON, 1974

Lake Malawi: Likoma, 8 cm

Photo: E.Schraml

A39960-5 *Labidochromis freibergi* JOHNSON, 1974

Lake Malawi: Likoma, 8 cm ♂

Photo: E.Schraml

A39960-5 *Labidochromis freibergi* JOHNSON, 1974

Lake Malawi: Likoma, 8 cm ♀

Photo: E.Schraml

A39995-5 *Labidochromis* spec. "Hongi"

Lake Malawi: Tanzanian Coast, 8-9 cm

Photo: Ch.Wyrwich

A40000-5 *Labidochromis* spec. "Hongi"

Lake Malawi: Hongi Isl. (Tanzania), 8-9 cm

Photo: A.Spreinat

A40000-5 *Labidochromis* spec. "Hongi"

Lake Malawi: Hongi Isl. (Tanzania), 8-9 cm

Photo: A.Spreinat

A40000-5 *Labidochromis* spec. "Hongi"

Lake Malawi: Hongi Isl. (Tanzania), 8-9 cm

Photo: A.Spreinat

A40000-5 *Labidochromis* spec. "Hongi"

Lake Malawi: Tanzanian Coast, 8-9 cm

Photo: M.K.Meyer

A40000-5 *Labidochromis* spec. "Hongi"

Lake Malawi: Tanzanian Coast, 8-9 cm

Photo: M.K.Meyer

A40000-5 *Labidochromis* spec. "Hongi"

Lake Malawi: Hongi Isl. (Tanzania), 8-9 cm

Photo: A.Spreinat

A39997-6 *Labidochromis* spec. "Hongi"

Lake Malawi: Hongi Isl. (Tanzania), 8-9 cm

Photo: W.Deproost

A40000-5 *Labidochromis* spec. "Hongi"

Lake Malawi: Hongi Isl. (Tanzania), 8-9 cm

Photo: A.Spreinat

A39996-5 *Labidochromis* spec. "Hongi"
Kimpuma-Labidochromis
Lake Malawi: Tanzanian Coast, 8-9 cm

Photo: M.K.Meyer

A40076-5 *Labidochromis* spec. "Mbamba"

Lake Malawi: Tanzanian Coast, 8-9 cm
♂ Photo: M.K.Meyer

A40076-5 *Labidochromis* spec. "Mbamba"

Lake Malawi: Tanzanian Coast, 8-9 cm
♀ Photo: M.K.Meyer

A40075-5 *Labidochromis* spec. "Mbamba"

Lake Malawi: Mbamba Bay (Tanzania): Ngkuyo Isl., 9 cm
♂ Photo: E.Schraml

A40075-5 *Labidochromis* spec. "Mbamba"

Lake Malawi: Mbamba Bay (Tanzania): Ngkuyo Isl., 9 cm
♀ Photo: M.K.Meyer

A40075-5 *Labidochromis* spec. "Mbamba"

Lake Malawi: Mbamba Bay (Tanzania): Ngkuyo Isl., 9 cm
♂ Photo: A.Spreinat

A40075-5 *Labidochromis* spec. "Mbamba"

Lake Malawi: Mbamba Bay (Tanzania): Ngkuyo Isl., 9 cm
♂ Photo: A.Spreinat

A40075-5 *Labidochromis* spec. "Mbamba"

Lake Malawi: Mbamba Bay (Tanzania), 9 cm
♂ Photo: L. Seegers

A40075-5 *Labidochromis* spec. "Mbamba"

Lake Malawi: Mbamba Bay (Tanzania), 9 cm
♀ Photo: L. Seegers

A39911-5 *Labidochromis* spec. "Blue White"
=? *Labidochromis caeruleus* Var.
Lake Malawi: Tumbi Reef (Tanzania), 9 cm
♂ Photo: A.Spreinat

A39911-5 *Labidochromis* spec. "Blue White"
=? *Labidochromis caeruleus* Var.
Lake Malawi: Tumbi Reef (Tanzania), 9 cm
♀ Photo: A.Spreinat

A40090-3 *Labidochromis mylodon* Lewis, 1982

Lake Malawi: Mumbo Isl., 9 cm
♂ Photo: Ad Konings

A40095-5 *Labidochromis pallidus* Lewis, 1982
spawning pair
Lake Malawi: Maleri Isl., 8 cm
♂ ♀ Photo: E.Schraml

A40095-5 *Labidochromis pallidus* Lewis, 1982

Lake Malawi: Maleri Isl., 8 cm
♂ Photo: E.Schraml

A40095-5 *Labidochromis pallidus* Lewis, 1982

Lake Malawi: Maleri Isl., 8 cm
♀ Photo: W.Deproost

A40095-5 *Labidochromis pallidus* Lewis, 1982

Lake Malawi: Nakanthenga Isl., 8 cm
♂ Photo: A.Spreinat

A40095-5 *Labidochromis pallidus* Lewis, 1982

Lake Malawi: Nakanthenga Isl., 8 cm
♀ Photo: A.Spreinat

A39905-5 *Labidochromis caeruleus* FRYER, 1956

Lake Malawi: Nkhata Bay, 8 cm
♂ Photo: A.Spreinat

A39905-5 *Labidochromis caeruleus* FRYER, 1956

Lake Malawi: Nkhata Bay, 8 cm
♀ Photo: A.Spreinat

A39913-5 *Labidochromis caeruleus* 'Yellow' FRYER, 1956
Yellow
Lake Malawi: Between Charo and Mbowe Isl., 13 cm
♂ Photo: E.Schraml

A39913-5 *Labidochromis caeruleus* 'Yellow' FRYER, 1956
Yellow
Lake Malawi: Between Charo and Mbowe Isl., 13 cm
♀ Photo: E.Schraml

A39907-5 *Labidochromis caeruleus* 'Blue' FRYER, 1956

Lake Malawi: Nkhata Bay, 13 cm
♂ Photo: Freds

A39908-5 *Labidochromis* spec. 'Black Dorsal'
= *Labidochromis caeruleus* 'Black Dorsal' FRYER, 1956 leg. KONINGS
Lake Malawi: Nkali (Tanzania), 10 cm
♂ Photo: E.Schraml

A39907-5 *Labidochromis* spec. 'Black Dorsal'
= *Labidochromis caeruleus* 'Black Dorsal' FRYER, 1956 leg. KONINGS
Lake Malawi: Lundo Isl. (Tanzania), 10 cm
♂ Photo: A.Spreinat

A39907-5 *Labidochromis* spec. 'Black Dorsal'
= *Labidochromis caeruleus* 'Black Dorsal' FRYER, 1956 leg. KONINGS
Lake Malawi: Lundo Isl. (Tanzania), 10 cm
♀ Photo: A.Spreinat

A39908-5 *Labidochromis* spec. 'Black Dorsal'
= *Labidochromis caeruleus* 'Black Dorsal' Fryer, 1956 leg. Konings
Lake Malawi: Nkali (Tanzania), 10 cm
♂ Photo: M.K.Meyer

A39912-5 *Labidochromis* spec. "Caeruleus Chilucha"
Lake Malawi: Chilucha Reef (Mozambique), 8 cm
♂ Photo: Ad Konings

A39940-5 *Labidochromis* spec. "Chisumulae Mbweca"
Lake Malawi: Tumbi Point (Mozambique), 9 cm
♂ Photo: Ad Konings

A39940-5 *Labidochromis* spec. "Chisumulae Mbweca"
Lake Malawi: Mbweca (Mozambique), 9 cm
♂ Photo: Ad Konings

A39935-5 *Labidochromis chisumulae* Lewis, 1982
Lake Malawi: Chisumulu, 8 cm
♂ Photo: W.Deproost

A39935-5 *Labidochromis chisumulae* Lewis, 1982
Lake Malawi: Chisumulu, 8 cm
♀ Photo: W.Deproost

A39935-5 *Labidochromis chisumulae* Lewis, 1982
Lake Malawi: Chisumulu, 8 cm
♂ Photo: W.Staeck

A39935-5 *Labidochromis chisumulae* Lewis, 1982
Lake Malawi: Chisumulu, 8 cm
♂ Photo: A.Spreinat

A39935-5 *Labidochromis chisumulae* Lewis, 1982

Lake Malawi: Chisumulu, 8 cm
♀ Photo: A.Spreinat

A39935-5 *Labidochromis chisumulae* Lewis, 1982

Lake Malawi: Chisumulu: Chilanje, 8 cm
♂ Photo: A.Spreinat

A39900-5 *Labidochromis* spec. "Blue Mapango"

Lake Malawi: Tanzanian Coast, 9 cm
♂ Photo: M.K.Meyer

A40145-5 *Labidochromis* spec. "Nose"

Lake Malawi: Tanzanian Coast, 9 cm
♂ Photo: M.K.Meyer

A40145-5 *Labidochromis* spec. "Nose"

Lake Malawi: Tanzanian Coast, 9 cm
♂ Photo: M.K.Meyer

A40145-5 *Labidochromis* spec. "Nose"

Lake Malawi: Tanzanian Coast, 9 cm
♀ Photo: M.K.Meyer

A40100-5 *Labidochromis* spec. "Perlmutt"

Lake Malawi: Mbamba Bay (Tanzania), 8-9 cm
♂ Photo: M.K.Meyer

A40100-5 *Labidochromis* spec. "Perlmutt"

Lake Malawi: Mbamba Bay (Tanzania), 8-9 cm
♀ Photo: M.K.Meyer

A40100-5 *Labidochromis* spec. "Perlmutt"

Lake Malawi: Mbamba Bay (Tanzania), 8-9 cm

Photo: A.Spreinat

A40100-5 *Labidochromis* spec. "Perlmutt"

Lake Malawi: Mbamba Bay (Tanzania): Higga Reef, 8-9 cm

Photo: A.Spreinat

A40025-4 *Labidochromis lividus* Lewis, 1982

Lake Malawi: Ndomo Point/Likoma, 6,5 cm

Photo: A. Spreinat

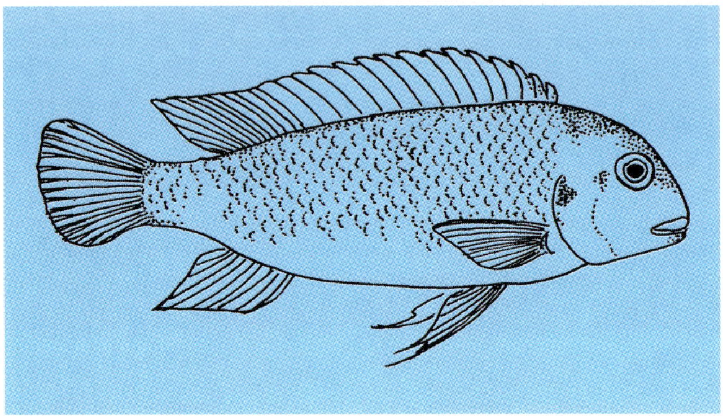

A40070 *Labidochromis mathotho* Burgess & Axelrod, 1976

Lake Malawi: exact place of origin unknown, 8 cm

Zeichnung: H. Nakano/A.C.S.

Eine typische Felsenküste am Malawisee
A typical rocky coast on the Lake Malawi shoreline

Photo: W. Staeck

A43850-5 *Melanochromis auratus* (Boulenger, 1897)
Türkisgoldbarsch / Nyasa Golden Cichlid
Lake Malawi: Mumbo Isl., 11 cm
♂ ♀ Photo: A.Spreinat

A43850-5 *Melanochromis auratus* (Boulenger, 1897)
Türkisgoldbarsch / Nyasa Golden Cichlid
Lake Malawi: SW-Coast (Jalo Reef to Crocodile Rock), 11 cm
♂ Photo: Migge-Reinhard / Archiv A.C.S.

A43850-5 *Melanochromis auratus* (Boulenger, 1897)
Türkisgoldbarsch / Nyasa Golden Cichlid
Lake Malawi: Domwe Isl., 11 cm
♂ Photo: W.Deproost

A43850-5 *Melanochromis auratus* (Boulenger, 1897)
Türkisgoldbarsch / Nyasa Golden Cichlid
Lake Malawi: SW-Coast (Jalo Reef to Crocodile Rock), 11 cm
♂ Photo: Ch.Wyrwich

A43850-5 *Melanochromis auratus* (Boulenger, 1897)
Türkisgoldbarsch / Nyasa Golden Cichlid
Lake Malawi: SW-Coast (Jalo Reef to Crocodile Rock), 11 cm
♂ Photo: E.Schraml

A43850-5 *Melanochromis auratus* (Boulenger, 1897)
Türkisgoldbarsch / Nyasa Golden Cichlid
Lake Malawi: SW-Coast (Jalo Reef to Crocodile Rock), 11 cm
♂ Photo: E.Schraml

A43850-4 *Melanochromis auratus* (Boulenger, 1897)
Türkisgoldbarsch / Nyasa Golden Cichlid
Lake Malawi: SW-Coast (Jalo Reef to Crocodile Rock), 11 cm
♀ Photo: E.Schraml

A43850-4 *Melanochromis auratus* (Boulenger, 1897)
mouthbreeding/maulbrütend
Lake Malawi: SW-Coast (Jalo Reef to Crocodile Rock), 11 cm
♀ Photo: E.Schraml

A43850-5 *Melanochromis auratus* (Boulenger, 1897)
mouthbreeding/maulbrütend, Männchenfärbung/male-coloration
Lake Malawi: SW-Coast (Jalo Reef to Crocodile Rock), 11 cm
Photo: E.Schraml

A43855-5 *Melanochromis* spec. "Auratus Elongate"

Lake Malawi: Miluluka: N'uosi Reef (Mozambique), 11 cm
Photo: A.Spreinat

A43855-5 *Melanochromis* spec. "Auratus Elongate"

Lake Malawi: Metangula: Chuanga (Mozambique), 11 cm
Photo: A.Spreinat

A43855-5 *Melanochromis* spec. "Auratus Elongate"

Lake Malawi: Chilucha Reef (Mozambique), 11 cm
Photo: Ad Konings

A43855-5 *Melanochromis* spec. "Auratus Elongate"

Lake Malawi: Minos Reef (Mozambique), 11 cm
Photo: Ad Konings

A43855-5 *Melanochromis* spec. "Auratus Elongate"

Lake Malawi: Minos Reef (Mozambique), 11 cm
Photo: Ad Konings

A43895-5 *Melanochromis dialeptos* Bowers & Stauffer, 1997
Zwerg-Auratus / Dwarf Auratus
Lake Malawi: SE-Coast at Masinje, 8 cm
Photo: E.Schraml

A43895-5 *Melanochromis dialeptos* Bowers & Stauffer, 1997
Zwerg-Auratus / Dwarf Auratus
Lake Malawi: SE-Coast at Masinje, 8 cm
Photo: W.Deproost

A43895-5 *Melanochromis dialeptos* BOWERS & STAUFFER, 1997
Zwerg-Auratus / Dwarf Auratus
Lake Malawi: Meponda: Nyamizimu (Mozambique), 8 cm
♂ Photo: A.Spreinat

A43895-5 *Melanochromis dialeptos* BOWERS & STAUFFER, 1997
Zwerg-Auratus / Dwarf Auratus
Lake Malawi: Meponda: Nyamizimu (Mozambique), 8 cm
♀ Photo: A.Spreinat

A43895-5 *Melanochromis dialeptos* BOWERS & STAUFFER, 1997
Zwerg-Auratus / Dwarf Auratus
Lake Malawi: Chikala (Mozambique), 8 cm
♂ Photo: A.Spreinat

A43895-5 *Melanochromis dialeptos* BOWERS & STAUFFER, 1997
Zwerg-Auratus / Dwarf Auratus
Lake Malawi: Chikala (Mozambique), 8 cm
♀ Photo: A.Spreinat

A43970-5 *Melanochromis simulans* ECCLES, 1973

Lake Malawi: E-Coast at Makanjila, 12 cm
♂ Photo: A.Spreinat

A43970-5 *Melanochromis simulans* ECCLES, 1973

Lake Malawi: E-Coast at Makanjila, 12 cm
♀ Photo: A.Spreinat

A43975-5 *Melanochromis vermivorus* TREWAVAS, 1935

Lake Malawi: Chinyamwezi, 15 cm
♂ Photo: Ad Konings

A43975-5 *Melanochromis vermivorus* TREWAVAS, 1935

Lake Malawi: Chinyamwezi, 15 cm
♀ Photo: Ad Konings

A43975-5 *Melanochromis vermivorus* TREWAVAS, 1935

Lake Malawi: Chinyankwazi, 15 cm

Photo: Ad Konings

A43975-5 *Melanochromis vermivorus* TREWAVAS, 1935

Lake Malawi: Chinyankwazi, 15 cm

Photo: A.Spreinat

A43975-5 *Melanochromis vermivorus* TREWAVAS, 1935

Lake Malawi: Mumbo Isl., 15 cm

Photo: A.Spreinat

A43975-5 *Melanochromis vermivorus* TREWAVAS, 1935
Yellow Form

Lake Malawi: Mumbo Isl., 15 cm

Photo: Ad Konings

A43975-5 *Melanochromis vermivorus* TREWAVAS, 1935

Lake Malawi: Maleri Isl., 15 cm

Photo: A.Spreinat

A43975-5 *Melanochromis vermivorus* TREWAVAS, 1935

Lake Malawi: Jalo Reef, 15 cm

Photo: Ad Konings

A43935-5 *Melanochromis* spec. 'Matema'

Lake Malawi: Matema (Tanzania), 10 cm

Photo: L. Seegers

Melanochromis chipokae JOHNSON, 1975

Lake Malawi: Chipoka, 15 cm

Photo: D.S. Johnson

A43950-5 *Melanochromis parallelus* Burgess & Axelrod, 1976
presumed to be a synonym of *M. chipokae*
Lake Malawi: Tanzanian Coast, 12 cm
♂ Photo: A.Spreinat

A43950-5 *Melanochromis parallelus* Burgess & Axelrod, 1976
presumed to be a synonym of *M. chipokae*
Lake Malawi: Tanzanian Coast, 12 cm
♀ Photo: A.Spreinat

A43950-5 *Melanochromis parallelus* Burgess & Axelrod, 1976
presumed to be a Synonym of *M. chipokae*
Lake Malawi, 12 cm
♂ Photo: A.Spreinat

A43950-5 *Melanochromis parallelus* Burgess & Axelrod, 1976
presumed to be a Synonym of *M. chipokae*
Lake Malawi, 12 cm
♂ Photo: E.Schraml

A43950-5 *Melanochromis parallelus* Burgess & Axelrod, 1976
presumed to be a synonym of *M. chipokae*
Lake Malawi, 12 cm
♂ Photo: W.Deproost

A43950-5 *Melanochromis parallelus* Burgess & Axelrod, 1976
presumed to be a synonym of *M. chipokae*
Lake Malawi: Nkhata Bay, 12 cm
♀ Photo: W.Deproost

A43950-5 *Melanochromis parallelus* Burgess & Axelrod, 1976
presumed to be a synonym of *M. chipokae*
Lake Malawi: Thumbi West Island, 12 cm
♂ Photo: A.Spreinat

A43950-5 *Melanochromis parallelus* Burgess & Axelrod, 1976
presumed to be a synonym of *M. chipokae*
Lake Malawi: Cobue (Mozambique): Namisi Rock, 12 cm
♂ Photo: A.Spreinat

A43950-5 *Melanochromis parallelus* Burgess & Axelrod, 1976
presumed to be a synonym of *M. chipokae*
Lake Malawi: Chisumulu: Membe Isl., 12 cm
Photo: A.Spreinat

A43950-5 *Melanochromis parallelus* Burgess & Axelrod, 1976
presumed to be a synonym of *M. chipokae*
Lake Malawi: Ponta Messuli (Mozambique), 12 cm
Photo: A.Spreinat

A43940-5 *Melanochromis melanopterus* Trewavas, 1935

Lake Malawi: Likoma: Membe Point, 13 cm
Photo: A.Spreinat

A43940-5 *Melanochromis melanopterus* Trewavas, 1935

Lake Malawi: Likoma: Membe Point, 13 cm
Photo: Ad Konings

A43940-5 *Melanochromis melanopterus* Trewavas, 1935

Lake Malawi, 13 cm
Photo:

A43940-5 *Melanochromis melanopterus* Trewavas, 1935

Lake Malawi: Likoma: Makulawe Point, 13 cm
Photo: A.Spreinat

A43942-5 *Melanochromis melanopterus* Trewavas, 1935
"Slab"-Form
Lake Malawi: Mbenji Isl., 13 cm
Photo: Ad Konings

A43942-5 *Melanochromis melanopterus* Trewavas, 1935
"Slab"-Form
Lake Malawi: Mbenji Isl., 13 cm
Photo: A.Spreinat

A43944-5 *Melanochromis melanopterus* Trewavas, 1935

Lake Malawi: Undu Reef (Tanzania), 13 cm

Photo: Ad Konings

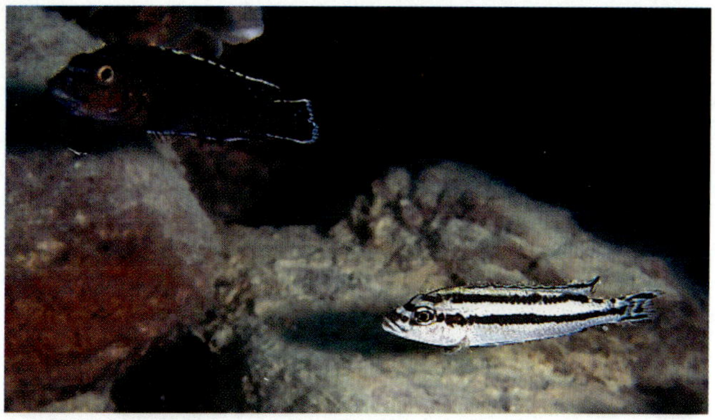

A43943-5 *Melanochromis* cf. *melanopterus* Trewavas, 1935
Females are looking like *M. simulans*
Lake Malawi: Likoma: Membe Point, 11 cm

Photo: A.Spreinat

A43943-5 *Melanochromis* cf. *melanopterus* Trewavas, 1935
Females are looking like *M. simulans*
Lake Malawi, 11 cm

Photo: E.Schraml

A43943-5 *Melanochromis* cf. *melanopterus* Trewavas, 1935
Females are looking like *M. simulans*
Lake Malawi, 11 cm

Photo: E.Schraml

A43932-5 *Melanochromis loriae* Johnson, 1975

Lake Malawi: Chidunga Rocks, 15 cm

Photo: Ad Konings

A43932-5 *Melanochromis loriae* Johnson, 1975

Lake Malawi: Chidunga Rocks, 15 cm

Photo: A.Spreinat

A43932-5 *Melanochromis loriae* Johnson, 1975

Lake Malawi: Chidunga Rocks, 15 cm

Photo: A.Spreinat

A43932-5 *Melanochromis loriae* Johnson, 1975

Lake Malawi: Nakantenga Reef, 15 cm

Photo: Ad Konings

A43932-5 *Melanochromis loriae* JOHNSON, 1975

Lake Malawi, 15 cm

Photo: Ch.Wyrwich

A43932-5 *Melanochromis loriae* JOHNSON, 1975

Lake Malawi, 15 cm

Photo: D.S.Johnson

A43886-5 *Melanochromis* cf. *chipokae* JOHNSON, 1975

Lake Malawi: Chipoka, 15 cm

Photo: E.Schraml

A43886-5 *Melanochromis* cf. *chipokae* JOHNSON, 1975

Lake Malawi: Chipoka, 15 cm

Photo: E.Schraml

A43860-5 *Melanochromis baliodigma* BOWERS & STAUFFER, 1997

Lake Malawi: Tumbi Point (Mozambique), 15 cm

Photo: Ad Konings

A43860-5 *Melanochromis baliodigma* BOWERS & STAUFFER, 1997

Lake Malawi: Tumbi Point (Mozambique), 15 cm

Photo: Ad Konings

A43860-5 *Melanochromis baliodigma* BOWERS & STAUFFER, 1997

Lake Malawi: Tumbi Point (Mozambique), 15 cm

Photo: A.Spreinat

A43860-5 *Melanochromis baliodigma* BOWERS & STAUFFER, 1997

Lake Malawi: Tumbi Point (Mozambique), 15 cm

Photo: A.Spreinat

A43985-4 *Melanochromis xanthodigma* BOWERS & STAUFFER, 1997

Lake Malawi: Masinje, 15 cm
♂ Photo: W.Staeck

A43985-4 *Melanochromis xanthodigma* BOWERS & STAUFFER, 1997

Lake Malawi: Meponda (Mozambique): Nyamizimu, 15 cm
♂ Photo: A.Spreinat

A43985-5 *Melanochromis xanthodigma* BOWERS & STAUFFER, 1997

Lake Malawi: Meponda (Mozambique): Nyamizimu, 15 cm
♂ Photo: A.Spreinat

A43985-5 *Melanochromis xanthodigma* BOWERS & STAUFFER, 1997

Lake Malawi: Meponda (Mozambique): Nyamizimu, 15 cm
♀ Photo: A.Spreinat

A43990-5 *Melanochromis* spec. "Northern Blue"

Lake Malawi: Undu (Tanzania), 15 cm
♂ Photo: M.K.Meyer

A43990-5 *Melanochromis* spec. "Northern Blue"

Lake Malawi: Lundu (Tanzania), 15 cm
♀ Photo: M.K.Meyer

A43990-5 *Melanochromis* spec. "Northern Blue"

Lake Malawi: Kirondo (Tanzania), 15 cm
♂ Photo: A.Spreinat

A43990-5 *Melanochromis* spec. "Northern Blue"

Lake Malawi: Lumbira (Tanzania), 15 cm
♀ Photo: A.Spreinat

A43866-5 *Melanochromis benetos* Bowers & Stauffer, 1997

Lake Malawi: Mbamba Bay (Tanzania): Chinula, 15 cm

♂ Photo: A.Spreinat

A43866-5 *Melanochromis benetos* Bowers & Stauffer, 1997

Lake Malawi: Mbamba Bay (Tanzania): Chinula, 15 cm

♀ Photo: A.Spreinat

A43960-5 *Melanochromis robustus* Johnson, 1985 leg. Konings

Lake Malawi: Chitande Isl., 15 cm

♂ Photo: Ad Konings

A43960-5 *Melanochromis robustus* Johnson, 1985 leg. Konings

Lake Malawi: Katale Isl., 15 cm

♂ Photo: Ad Konings

A43965-5 *Melanochromis* spec. "Robustus Mbenji"

Lake Malawi: Mbenji Isl., 15 cm

♂ Photo: Ad Konings

A43962-5 *Melanochromis robustus* Johnson, 1985

Lake Malawi: Likoma Isl., 15 cm

♂ Photo: D.S.Johnson

A43930-5 *Melanochromis lepidiadaptes* Bowers & Stauffer, 1997

Lake Malawi: Makanjila Point, 15 cm

♂ Photo: E Schraml

A43930-5 *Melanochromis lepidiadaptes* Bowers & Stauffer, 1997

Lake Malawi: Makanjila Point, 15 cm

♀ Photo: E.Schraml

A43920-5 *Melanochromis johannii* (Eccles, 1973)
Kobaltorangebarsch
Lake Malawi, 12 cm
♂ Photo: E.Schraml

A43920-5 *Melanochromis johannii* (Eccles, 1973)
Kobaltorangebarsch
Lake Malawi, 12 cm
♀ Photo: E.Schraml

A43920-5 *Melanochromis johannii* (Eccles, 1973)
Kobaltorangebarsch
Lake Malawi, 12 cm
♀ Photo: E.Schraml

A43920-4 *Melanochromis johannii* (Eccles, 1973)
Kobaltorangebarsch
Lake Malawi, 12 cm
♂ Photo: A.Spreinat

A43920-5 *Melanochromis johannii* (Eccles, 1973)
Kobaltorangebarsch
Lake Malawi: Makanjila, 12 cm
♂ Photo: A.Spreinat

A43920-5 *Melanochromis johannii* (Eccles, 1973)
Kobaltorangebarsch
Lake Malawi: Makanjila, Masinje, 12 cm
♂ Photo: A.Spreinat

A43920-5 *Melanochromis johannii* (Eccles, 1973)
Kobaltorangebarsch
Lake Malawi: Metangula (Mozambique): Lupuchi, 12 cm
♂ Photo: A.Spreinat

A43920-5 *Melanochromis johannii* (Eccles, 1973)
Kobaltorangebarsch
Lake Malawi: Metangula (Mozambique): Chuanga, 12 cm
♂ Photo: A.Spreinat

A43890-5 *Melanochromis cyaneorhabdos* Bowers & Stauffer, 1997

Lake Malawi: Likoma, 10 cm

♂ Photo: E.Schraml

A43890-5 *Melanochromis cyaneorhabdos* Bowers & Stauffer, 1997

Lake Malawi: Likoma, 10 cm

♀ Photo: E.Schraml

A43900-5 *Melanochromis interruptus* Johnson, 1975

Lake Malawi: SW-Chisumulu, 10 cm

♂ Photo: A.Spreinat

A43900-5 *Melanochromis interruptus* Johnson, 1975

Lake Malawi: SW-Chisumulu, 10 cm

♀ Photo: A.Spreinat

A43900-5 *Melanochromis interruptus* Johnson, 1975

Lake Malawi: SW-Chisumulu, 10 cm

♂ Photo: A.Spreinat

A43900-5 *Melanochromis interruptus* Johnson, 1975

Lake Malawi: SW-Chisumulu, 10 cm

♀ Photo: A.Spreinat

A43955-5 *Melanochromis perileucos* Bowers & Stauffer, 1997

Black & White Johanni

Lake Malawi: Likoma, 10 cm

♂ Photo: A.Spreinat

A43955-5 *Melanochromis perileucos* Bowers & Stauffer, 1997

Black & White Johanni, 2 males fighting

Lake Malawi: Likoma: Mbamba Isl., 10 cm

♂ Photo: W.Deproost

A43955-5 *Melanochromis perileucos* BOWERS & STAUFFER, 1997
Black & White Johanni, mouthbreeding
Lake Malawi: Likoma: Msekwa Point, 10 cm
Photo: A.Spreinat

A43955-5 *Melanochromis perileucos* BOWERS & STAUFFER, 1997
Black & White Johanni
Lake Malawi: Likoma: Khuyu, 10 cm
Photo: A.Spreinat

A43915-5 *Melanochromis joanjohnsonae* (JOHNSON, 1974)
Perle von Likoma
Lake Malawi: Likoma, 10 cm
Photo: A.Spreinat

A43915-4 *Melanochromis joanjohnsonae* (JOHNSON, 1974)
Perle von Likoma
Lake Malawi: Thumbi West Isl., 10 cm
Photo: A.Spreinat

A43915-4 *Melanochromis joanjohnsonae* (JOHNSON, 1974)
Perle von Likoma
Lake Malawi: Likoma: Khuyu, 10 cm
Photo: A.Spreinat

A43915-3 *Melanochromis joanjohnsonae* (JOHNSON, 1974)
Perle von Likoma
Lake Malawi: Likoma, 10 cm
Photo: E.Schraml

A43915-4 *Melanochromis joanjohnsonae* (JOHNSON, 1974)
Perle von Likoma
Lake Malawi: Likoma, 10 cm
Photo: D.S.Johnson

A43995-5 *Melanochromis* spec. 'Tumbi'

Lake Malawi: Tumbi, >12 cm
Photo: M.K.Meyer

A43875-5 *Melanochromis* cf. *brevis* TREWAVAS, 1935

Lake Malawi: Chinyankwazi, 12 cm

Photo: Ad Konings

A43875-5 *Melanochromis* cf. *brevis* TREWAVAS, 1935

Lake Malawi, 12 cm

Photo: W.Staeck

A43880-5 *Melanochromis* spec. "Brown"

Lake Malawi: Boadzulu, 13 cm

Photo: A. Spreinat

A43925-5 *Melanochromis labrosus* TREWAVAS, 1935

Lake Malawi: Tanzanian Coast, 13 cm

Photo: M.K.Meyer

A43925-5 *Melanochromis labrosus* TREWAVAS, 1935

Lake Malawi: Mbamba Bay (Tanzania), 13 cm

Photo: A.Spreinat

A43925-4 *Melanochromis labrosus* TREWAVAS, 1935

Lake Malawi: Mbamba Bay (Tanzania), 13 cm

Photo: A.Spreinat

A43925-4 *Melanochromis labrosus* TREWAVAS, 1935

Lake Malawi: Malawian S-Coast, 13 cm

Photo: E.Schraml

A43925-5 *Melanochromis labrosus* TREWAVAS, 1935

Lake Malawi, 13 cm

Photo: E.Schraml

A56800-5 *Petrotilapia* spec. "Black Flank"

Lake Malawi: Chitendi Isl., ? cm

♂ Photo: A.Spreinat

A56800-5 *Petrotilapia* spec. "Black Flank"

Lake Malawi: Mpanga Rocks, ? cm

♂ Photo: A.Spreinat

A56865-5 *Petrotilapia nigra* MARSH, 1983

Lake Malawi: Thumbi West, 14 cm

♂ Photo: A.Spreinat

A56865-5 *Petrotilapia nigra* MARSH, 1983

Lake Malawi: Thumbi West, 14 cm

♂ Photo: A.Spreinat

A56865-5 *Petrotilapia nigra* MARSH, 1983

Lake Malawi: Otter Point, 14 cm

♂ Photo: Ad Konings

A56870-5 *Petrotilapia* spec. "Nigra Tumbi"

Lake Malawi: Tumbi Point (Mozambique), 14 cm

♂ Photo: Ad Konings

A56870-5 *Petrotilapia spec.* "Nigra Tumbi"

Lake Malawi: Cobue (Mozambique): Mara Point, 14 cm

♂ Photo: A.Spreinat

A56870-5 *Petrotilapia* spec. "Nigra Tumbi"

Lake Malawi: Tumbi Point (Mozambique), 14 cm

♀ Photo: A.Spreinat

A56907-5 *Petrotilapia* spec. "Tanzania"

Lake Malawi: Kirondo (Tanzania), 17 cm

Photo: A.Spreinat

A56913-5 *Petrotilapia* spec. "Tanzania"

Lake Malawi: Njambe Bay (N) (Tanzania), 17 cm

Photo: A.Spreinat

A56911-5 *Petrotilapia* spec. "Tanzania"

Lake Malawi: Mbamba Bay: Ngkuyo Isl. (Tanzania), 17 cm

Photo: A.Spreinat

A56908-5 *Petrotilapia* spec. "Tanzania"

Lake Malawi: Magunga (Tanzania), 17 cm

Photo: Ad Konings

A56914-5 *Petrotilapia* spec. "Tanzania"

Lake Malawi: Makonde (Tanzania), 17 cm

Photo: Ad Konings

A56915-5 *Petrotilapia* spec. "Tanzania"

Lake Malawi: Nkanda (Tanzania), 17 cm

Photo: A.Spreinat

A56855-5 *Petrotilapia* spec. "Mumbo Blue"

Lake Malalwi: Mumbo Isl., 17 cm

Photo: A.Spreinat

A56845-5 *Petrotilpia* spec. "Likoma Barred"

Lake Malawi: Likoma: Mazimbwe Isl., 17 cm

Photo: A.Spreinat

A56850-5 *Petrotilapia* spec. "Likoma Variable"

Lake Malawi: Likoma: Mazimbwe Isl., 17 cm
♂ Photo: A.Spreinat

A56850-5 *Petrotilapia* spec. "Likoma Variable"

Lake Malawi: Likoma: Mazimbwe Isl., 17 cm
♀ Photo: A.Spreinat

A56830-5 *Petrotilapia genalutea* MARSH, 1983

Lake Malawi: Mundola, 16 cm
♂ Photo: Ad Konings

A56830-5 *Petrotilapia genalutea* MARSH, 1983

Lake Malawi: Thumbi West Isl., 16 cm
♀ Photo: A.Spreinat

A56830-5 *Petrotilapia genalutea* MARSH, 1983

Lake Malawi: Nakanthenga, 16 cm
♂ Photo: A.Spreinat

A56830-5 *Petrotilapia genalutea* MARSH, 1983

Lake Malawi: Pombo Rocks (Tanzania), 16 cm
♂ Photo: Ad Konings

A56945-5 *Petrotilapia* spec. "Pointed Head"
=? *Petrotilapia genalutea*
Lake Malawi: Hai Reef (Tanzania), 14-16 cm
♂ Photo: A.Spreinat

A56945-5 *Petrotilapia* spec. "Pointed Head"
=? *Petrotilapia genalutea*; Foto: School of females
Lake Malawi: Hai Reef (Tanzania), 14-16 cm
♀ Photo: A.Spreinat

A56815-5 *Petrotilapia chrysos* Stauffer & Van Snik,1996

Lake Malawi: Chinyamwezi Isl., 17 cm
♂ Photo: A.Spreinat

A56815-5 *Petrotilapia chrysos* Stauffer & Van Snik,1996

Lake Malawi: Chinyamwezi Isl., 17 cm
♂ Photo: A.Spreinat

A56815-5 *Petrotilapia chrysos* Stauffer & Van Snik,1996

Lake Malawi: Chinyamwezi Isl., 17 cm
♂ Photo: A.Spreinat

A56815-5 *Petrotilapia chrysos* Stauffer & Van Snik,1996

Lake Malawi: Chinyamwezi Isl., 17 cm
♀ Photo: A.Spreinat

A56835-5 *Petrotilapia* spec. "Gold Eastern"

Lake Malawi: Nyamizimu/Meponda (Mozambique), 15 cm
♂ Photo: A.Spreinat

A56835-5 *Petrotilapia* spec. "Gold Eastern"

Lake Malawi: Gome Rock/ Fort Maguire, 15 cm
♂ Photo: A.Spreinat

A56835-5 *Petrotilapia* spec. "Gold Eastern"

Lake Malawi: Gome Rock/ Fort Maguire, 15 cm
♀ Photo: A.Spreinat

A56835-4 *Petrotilapia* spec. "Gold Eastern"

Lake Malawi: Gome, 15 cm
♀ Photo: Ad Konings

A56875-5 *Petrotilapia* spec. "Nigra Yellow Dorsal"

Lake Malawi: Chiloelo (Mozambique), ? cm
♂ Photo: Ad Konings

A56875-5 *Petrotilapia* spec. "Nigra Yellow Dorsal"

Lake Malawi: Chiloelo (Mozambique), ? cm
♀ Photo: Ad Konings

A56937-5 *Petrotilapia* spec. "Yellow Ventral"

Lake Malawi: Londo (Mozambique), ? cm
♂ Photo: Ad Konings

A56936-5 *Petrotilapia* spec. "Yellow Ventral"

Lake Malawi: Chilanje/Chisumulu, ? cm
♂ Photo: A.Spreinat

A56939-5 *Petrotilapia* spec. "Yellow Ventral"

Lake Malawi: Hai Reef (Tanzania), ? cm
♂ Photo: A.Spreinat

A56938-5 *Petrotilapia* spec. "Yellow Ventral"

Lake Malawi: Lumbaulo (Mozambique), ? cm
♂ Photo: Ad Konings

A56940-5 *Petrotilapia* spec. "Yellow Ventral"

Lake Malawi: Ponta Messuli (Mozambique), ? cm
♂ Photo: A.Spreinat

A56940-5 *Petrotilapia* spec. "Yellow Ventral"

Lake Malawi: Ponta Messuli (Mozambique), ? cm
♀ Photo: A.Spreinat

A56936-5 *Petrotilapia* spec. "Yellow Ventral"

Lake Malawi: Chiwe Rocks/Chisumulu, ? cm

Photo: A.Spreinat

A56941-5 *Petrotilapia* spec. "Yellow Ventral"

Lake Malawi: Kanjindo/Cobue (Mozambique), ? cm

Photo: A.Spreinat

A56810-5 *Petrotilapia* spec. "Chitimba"

Lake Malawi: Chitimba Bay, ? cm

Photo: Ad Konings

A56810-2 *Petrotilapia* spec. "Chitimba"

Lake Malawi: Chitimba Bay, ? cm

Photo: Ad Konings

A56825-5 *Petrotilapia* spec. "Fuscous"

Lake Malawi: Maleri Isl., 17 cm

Photo: Ad Konings

A56825-5 *Petrotilapia* spec. "Fuscous"

Lake Malawi: Maleri Isl., 17 cm

Photo: Ad Konings

A56825-5 *Petrotilapia* spec. "Fuscous"

Lake Malawi: Nakanthenga Isl., 17 cm

Photo: A.Spreinat

A56891-5 *Petrotilapia* spec. "Ruarwe"

Lake Malawi: Charo, 17 cm

Photo: Ad Konings

A56892-5 *Petrotilapia* spec. "Ruarwe"

Lake Malawi: Chitendi/Chilumba, 17 cm
♂ Photo: A.Spreinat

A56892-5 *Petrotilapia* spec. "Ruarwe"

Lake Malawi: Mpanga Rocks/Chilumba, 17 cm
♂ Photo: A.Spreinat

A56892-5 *Petrotilapia* spec. "Ruarwe"

Lake Malawi: Chitendi/Chilumba, 17 cm
♀ Photo: A.Spreinat

A56895-4 *Petrotilapia* spec. "Ruarwe"

Lake Malawi: Kakusa, 17 cm
♀ Photo: Ad Konings

A56893-5 *Petrotilapia* spec. "Ruarwe"

Lake Malawi: Mbowe Isl., 17 cm
♂ Photo: Ad Konings

A56894-5 *Petrotilapia* spec. "Ruarwe"

Lake Malawi: Mdoka, 17 cm
♂ Photo: Ad Konings

A56900-5 *Petrotilapia* spec. "Small Blue"

Lake Malawi: Nkhata Bay, 14 cm
♂ Photo: A.Spreinat

A56900-5 *Petrotilapia* spec. "Small Blue"

Lake Malawi: Nkhata Bay, 14 cm
♀ Photo: A.Spreinat

A56900-5 *Petrotilapia* spec. "Small Blue"

Lake Malawi: Nkhata Bay, 14 cm

Photo: Ad Konings

A56880-5 *Petrotilapia* spec. "Orange Pelvic"

Lake Malawi: Chizumulu Isl., ? cm

Photo: Ad Konings

A56880-5 *Petrotilapia* spec. "Orange Pelvic"

Lake Malawi: Membe Isl./Chizumulu, ? cm

Photo: A.Spreinat

A56880-5 *Petrotilapia* spec. "Orange Pelvic"

Lake Malawi: Chizumulu Isl., ? cm

Photo: A.Spreinat

A56860-5 *Petrotilapia* spec. "Mumbo Yellow"

Lake Malawi: Mumbo Isl., 17 cm

Photo: A.Spreinat

A56860-5 *Petrotilapia* spec. "Mumbo Yellow"

Lake Malawi: Mumbo Isl., 17 cm

Photo: A.Spreinat

A56930-5 *Petrotilapia* spec. "Yellow Chin"

Lake Malawi: Maleri Isl., 17 cm

Photo: A.Spreinat

A56930-5 *Petrotilapia* spec. "Yellow Chin"

Lake Malawi: Maleri Isl., 17 cm

Photo: A.Spreinat

A56885-6 *Petrotilapia* spec. "Retrognathous"

Lake Malawi: Chizumulu, 19 cm
♂ Photo: Ad Konings

A56885-5 *Petrotilapia* spec. "Retrognathous"

Lake Malawi: Chizumulu, 19 cm
♀ Photo: Ad Konings

A56805-4 *Petrotilapia* spec. "Chitande"

Lake Malawi: Chitendi Isl./Chilumba, ? cm
♂ Photo: A.Spreinat

A56950-4 *Petrotilapia* spec.

Lake Malawi, ? cm
♂ Photo: E.Schraml

A56924-5 *Petrotilapia tridentiger* Trewavas, 1935

Lake Malawi: Hongi Isl. (Tanzania), 20 cm
♂ Photo: A.Spreinat

A56921-5 *Petrotilapia tridentiger* Trewavas, 1935

Lake Malawi: Boadzulu Isl., 20 cm
♂ Photo: A.Spreinat

A56922-5 *Petrotilapia tridentiger* Trewavas, 1935

Lake Malawi: Chia Lagoon, 20 cm
♂ Photo: Ad Konings

A56921-5 *Petrotilapia tridentiger* Trewavas, 1935

Lake Malawi: Boadzulu Isl., 20 cm
♀ Photo: Ad Konings

A78470-5 *Maylandia livingstonii* (BOULENGER, 1899)

Lake Malawi + Malambo, 6-13 cm

Photo: A.Spreinat

A78470-5 *Maylandia livingstonii* (BOULENGER, 1899)

Lake Malawi + Malambo, 6-13 cm

Photo: A.Spreinat

A78470-5 *Maylandia livingstonii* (BOULENGER, 1899)

Lake Malawi + Malambo, 6-13 cm

Photo: A.Spreinat

A78470-5 *Maylandia livingstonii* (BOULENGER, 1899)

Lake Malawi + Malambo, 6-13 cm

Photo: E.Schraml

A78470-5 *Maylandia livingstonii* (BOULENGER, 1899)

Lake Malawi + Malambo, 6-13 cm

Photo: E.Schraml

A78470-5 *Maylandia livingstonii* (BOULENGER, 1899)

Lake Malawi + Malambo, 6-13 cm

Photo: W.Staeck

A78472-5 *Maylandia livingstonii* (BOULENGER, 1899)

Lake Malawi: Crocodile Rock, 6-13 cm

Photo: Ad Konings

A78474-5 *Maylandia livingstonii* (BOULENGER, 1899)

Lake Malawi: Otter Point, 6-13 cm

Photo: Ad Konings

A78471-5 *Maylandia lanisticola* (Burgess, 1976)
presumed to be a synonym of *M. livingstonii*
Lake Malawi: Chesese, 6-13 cm
♀ Photo: Ad Konings

A78476-5 *Maylandia lanisticola* (Burgess, 1976)
presumed to be a synonym of *M. livingstonii*
Lake Malawi: Eccles Reef, 6-13 cm
♂ Photo: A.Spreinat

A78477-5 *Maylandia lanisticola* (Burgess, 1976)
presumed to be a synonym of *M. livingstonii*
Lake Malawi: 6-13 cm
♂ Photo: E.Schraml

A78477-5 *Maylandia lanisticola* (Burgess, 1976)
presumed to be a synonym of *M. livingstonii*
Lake Malawi: 6-13 cm
♀ Photo: E.Schraml

A78477-5 *Maylandia lanisticola* (Burgess, 1976)
presumed to be a synonym of *M. livingstonii*
Lake Malawi: 6-13 cm
♀ Photo: Ch. Wyrwich

A78654-5 *Maylandia pursa* (Stauffer, 1991)
presumed to be a synonym of *M. livingstonii*
Lake Malawi: Cape Maclear area, 6-13 cm
♂ Drawing: H. Nakano/A.C.S.

A78473-5 *Maylandia* spec. "Livingstonii Likoma"

Lake Malawi: Mbamba Bay (Tanzania): Chinula
♂ Photo: A.Spreinat

A78473-5 *Maylandia* spec. "Livingstonii Likoma"

Lake Malawi: Mbahwa Isl. (Tanzania)
♀ Photo: A.Spreinat

A78475-5 *Maylandia* spec. "Livingstonii Likoma"

Lake Malawi, 6-13 cm

Photo: E.Schraml

A78473-5 *Maylandia* spec. "Livingstonii Likoma"

Lake Malawi: Likoma: Khuyu,

Photo: A.Spreinat

A78195-5 *Maylandia elegans* (Trewavas, 1935)

Lake Malawi: Lakewide distribution, 15 cm

Photo: A.Spreinat

A78195-4 *Maylandia elegans* (Trewavas, 1935)

Lake Malawi: Senga Bay, 15 cm

Photo: A.Spreinat

A78480-6 *Maylandia lombardoi* (Burgess, 1977)

Lake Malawi: Mbenji Isl., Nkhomo Reef, 15 cm

Photo: E.Schraml

A78480-5 *Maylandia lombardoi* (Burgess, 1977)

Lake Malawi: Mbenji Isl., Nkhomo Reef, 15 cm

Photo: E.Schraml

A78480-4 *Maylandia lombardoi* (Burgess, 1977)

Lake Malawi: Mbenji Isl., Nkhomo Reef, 15 cm

Photo: E.Schraml

A78480-2 *Maylandia lombardoi* (Burgess, 1977)

Lake Malawi: Mbenji Isl., Nkhomo Reef, 15 cm

Photo: Nakano/A.C.S.

A78480-6 *Maylandia lombardoi* (Burgess, 1977)

Lake Malawi: Mbenji Isl., Nkhomo Reef, 15 cm

♀ Photo: A.Spreinat

A78480-5 *Maylandia lombardoi* (Burgess, 1977)

Lake Malawi: Mbenji Isl., Nkhomo Reef, 15 cm

♀ Photo: Ch.Wyrwich

A78065-6 *Maylandia barlowi* (McKaye & Stauffer, 1986)

Lake Malawi: Mbenji Isl., 11 cm

♂ Photo: W.Deproost

A78065-3 *Maylandia barlowi* (McKaye & Stauffer, 1986)

Lake Malawi: Mbenji Isl., 11 cm

♀ Photo: A.Spreinat

A78065-5 *Maylandia barlowi* (McKaye & Stauffer, 1986)

Lake Malawi: Mbenji Isl., 11 cm

♂ Photo: E.Schraml

A78065-6 *Maylandia barlowi* (McKaye & Stauffer, 1986)

Lake Malawi: Nakanthenga, 11 cm

♂ Photo: A.Spreinat

A78420-6 *Maylandia hajomaylandi* (in 't Veen, 1984)

Lake Malawi: Chisumulu Isl., 12 cm

♂ Photo: E.Schraml

A78420-5 *Maylandia hajomaylandi* (in 't Veen, 1984)

Lake Malawi: Chisumulu Isl., 12 cm

♂ Photo: W.Staeck

A78420-5 *Maylandia hajomaylandi* (IN 'T VEEN, 1984)

Lake Malawi: Chisumulu Isl., 12 cm

♂ Photo: A.Spreinat

A78420-5 *Maylandia hajomaylandi* (IN 'T VEEN, 1984)

Lake Malawi: Chisumulu Isl., 12 cm

♀ Photo: A.Spreinat

A78415-6 *Maylandia greshakei* (MEYER & FÖRSTER, 1984)

Lake Malawi: Makokola Reef, 15 cm

♂ Photo: E.Schraml

A78415-5 *Maylandia greshakei* (MEYER & FÖRSTER, 1984)
maulbrütend
Lake Malawi: Makokola Reef, 15 cm

♀ Photo: E.Schraml

A78415-6 *Maylandia greshakei* (MEYER & FÖRSTER, 1984)

Lake Malawi: Makokola Reef, 15 cm

♂ Photo: E.Schraml

A78416-5 *Maylandia greshakei* X *Maylandia zebra*
OB-Morph
15 cm

♀ Photo: E.Schraml

A43625-4 *Maylandia mbenjii* (STAUFFER, BOWERS, KELLOG & MCKAYE, 1997)
Albino: Breeding form
Lake Malawi: 11 cm

♂ Photo: Taikong Corp.

A79225-5 *Maylandia mbenjii* (STAUFFER, BOWERS, KELLOG & MCKAYE, 1997)

Lake Malawi: Mbenji Isl., 11 cm

♂ Photo: E.Schraml

A79225-5 *Maylandia mbenjii* (Stauffer, Bowers, Kellog & McKaye, 1997)

Lake Malawi: Mbenji Isl., 11 cm

Photo: A.Spreinat

A79225-6 *Maylandia mbenjii* (Stauffer, Bowers, Kellog & McKaye, 1997)
OB-Morph
Lake Malawi: Mbenji Isl., 11 cm

Photo: W.Staeck

A79225-5 *Maylandia mbenjii* (Stauffer, Bowers, Kellog & McKaye, 1997)
OB-Morph
Lake Malawi: Mbenji Isl., 11 cm

Photo: A.Spreinat

A79225-4 *Maylandia mbenjii* (Stauffer, Bowers, Kellog & McKaye, 1997)
OB-Morph
Lake Malawi: Mbenji Isl., 11 cm

Photo: A.Spreinat

A79225-5 *Maylandia mbenjii* (Stauffer, Bowers, Kellog & McKaye, 1997)
O-Morph
Lake Malawi: Mbenji Isl., 11 cm

Photo: A.Spreinat

A79225-5 *Maylandia mbenjii* (Stauffer, Bowers, Kellog & McKaye, 1997)
O-Morph
Lake Malawi: Mbenji Isl., 11 cm

Photo: A.Spreinat

A78370-5 *Maylandia estherae* (Konings, 1995)

Lake Malawi: N'nosi Reef (Mozambique), 11-13 cm

Photo: A.Spreinat

A78370-5 *Maylandia estherae* (Konings, 1995)
O-Morph
Lake Malawi: N'nosi Reef (Mozambique), 11-13 cm

Photo: A.Spreinat

A78370-5 *Maylandia estherae* (Konings, 1995)

Lake Malawi: N'nosi Reef (Mozambique), 11-13 cm

♀ Photo: A.Spreinat

A78370-5 *Maylandia estherae* (Konings, 1995)
OB-Morph
Lake Malawi: N'nosi Reef (Mozambique), 11-13 cm

♀ Photo: A.Spreinat

A78370-5 *Maylandia estherae* (Konings, 1995)
O-Male (Wild)
Lake Malawi: N'nosi Reef (Mozambique), 11-13 cm

♂ Photo: A.Spreinat

A78370-6 *Maylandia estherae* (Konings, 1995)
O-Male (Breeding-Form)
Lake Malawi: N'nosi Reef (Mozambique), 11-13 cm

♂ Photo: W.Staeck

A78371-5 *Maylandia estherae* (Konings, 1995)

Lake Malawi: Ukhungu Reef (Mozambique), 11-13 cm

♂ Photo: A.Spreinat

A78371-5 *Maylandia estherae* (Konings, 1995)

Lake Malawi: Ukhungu Reef (Mozambique), 11-13 cm

♀ Photo: A.Spreinat

A78372-6 *Maylandia estherae* (Konings, 1995)
OB-Morph
Lake Malawi: Metangula (Mozambique), 11-13 cm

♂ Photo: W.Staeck

A78372-5 *Maylandia estherae* (Konings, 1995)
OB-Morph
Lake Malawi: Metangula (Mozambique), 11-13 cm

♂ Photo: E.Schraml

A78372-6 *Maylandia estherae* (Konings, 1995)
O-Morph
Lake Malawi: Metangula (Mozambique), 11-13 cm
Photo: A.Spreinat

A78372-5 *Maylandia estherae* (Konings, 1995)
O-Morph
Lake Malawi: Metangula (Mozambique), 11-13 cm
Photo: E.Schraml

A78372-5 *Maylandia estherae* (Konings, 1995)
O-Morph
Lake Malawi: Metangula (Mozambique), 11-13 cm
Photo: A.Spreinat

A78372-2 *Maylandia estherae* (Konings, 1995)
O-Morph
Lake Malawi: Metangula (Mozambique), 11-13 cm
Photo: Jap.

A78370-5 *Maylandia estherae* (Konings, 1995)
O-Morph
Lake Malawi: N'nosi Reef (Mozambique), 11-13 cm
Photo: A.Spreinat

A78372-5 *Maylandia estherae* (Konings, 1995)
OB-Morph
Lake Malawi: Metangula (Mozambique), 11-13 cm
♂ Photo: E.Schraml

A78372-5 *Maylandia estherae* (Konings, 1995)
OB-Morph
Lake Malawi: Metangula (Mozambique), 11-13 cm
♀ Photo: E.Schraml

A78375-5 *Maylandia* spec. "Estherae Blueface"

Lake Malawi: Lumessi North (Mozambique), 11-13 cm
♂ Photo: Ad Konings

A78375-5 *Maylandia* spec. "Estherae Blueface"
O-Morph
Lake Malawi: Lumessi North (Mozambique), 11-13 cm
♂ Photo: Ad Konings

A78375-5 *Maylandia* spec. "Estherae Blueface"

Lake Malawi: Chiloelo (Mozambique): Ngulu, 11-13 cm
♂ Photo: A.Spreinat

A78375-5 *Maylandia* spec. "Estherae Blueface"

Lake Malawi: Chikala (Mozambique), 11-13 cm
♂ Photo: A.Spreinat

A78375-5 *Maylandia* spec. "Estherae Blueface"

Lake Malawi: Chikala (Mozambique), 11-13 cm
♂ Photo: A.Spreinat

A78375-5 *Maylandia* spec. "Estherae Blueface"

Lake Malawi: Chikala (Mozambique), 11-13 cm
♀ Photo: A.Spreinat

A78133-5 *Maylandia callainos* (Stauffer & Hert, 1992)

Lake Malawi: Chilumba: Katari Isl., 11-13 cm

Photo: A.Spreinat

A78134-5 *Maylandia callainos* (Stauffer & Hert, 1992)

Lake Malawi: Chilumba: Chewere, 11-13 cm

Photo: A.Spreinat

A78132-5 *Maylandia callainos* (Stauffer & Hert, 1992)

Lake Malawi: Nkhata Bay, 11-13 cm

Photo: W.Deproost

A78136-5 *Maylandia callainos* (Stauffer & Hert, 1992)
OB-Morph
Lake Malawi: Lion's Cove, 11-13 cm

Photo: A.Spreinat

A78137-5 *Maylandia callainos* (Stauffer & Hert, 1992)
OB-Morph
Lake Malawi, 11-13 cm

Photo: A.Spreinat

A78135-5 *Maylandia callainos* (Stauffer & Hert, 1992)

Lake Malawi: Puulu (Tanzania), 11-13 cm

Photo: A.Spreinat

A78131-5 *Maylandia callainos* (Stauffer & Hert, 1992)
White-Morph
Lake Malawi, 11-13 cm

Photo: E.Schraml

A78131-5 *Maylandia callainos* (Stauffer & Hert, 1992)
White-Morph
Lake Malawi, 11-13 cm

Photo: E.Schraml

A78132-5 *Maylandia callainos* (STAUFFER & HERT, 1992)

Lake Malawi: Nkhata Bay, 11-13 cm

Photo: W.Deproost

A78132-4 *Maylandia callainos* (STAUFFER & HERT, 1992)

Lake Malawi: Nkhata Bay, 11-13 cm

Photo: A.Spreinat

A78130-5 *Maylandia callainos* (STAUFFER & HERT, 1992)

Lake Malawi: 11-13 cm

Photo: Ch.Wyrwich

A79370-5 *Maylandia benetos* (STAUFFER, BOWERS, KELLOG & MCKAYE, 1997)
Species not accepted by some authors
Lake Malawi: Mazinzi Reef, 11-13 cm

Photo: J.Stauffer

Unterwasseransicht von Chitendi Island (Chilumba)
Underwater impression of Chitendi Island (Chilumba)

Photo: A.Spreinat

A79235-5 *Maylandia* spec. "Zebra Mozambique"

Lake Malawi: Gome, 12 cm
♂ Photo: Ad Konings

A79235-4 *Maylandia* spec. "Zebra Mozambique"

Lake Malawi: Gome, 12 cm
♀ Photo: Ad Konings

A79236-5 *Maylandia* spec. "Zebra Mozambique"

Lake Malawi: Makanjila, 12 cm
♂ Photo: A.Spreinat

A79236-4 *Maylandia* spec. "Zebra Mozambique"

Lake Malawi: Ostküste, 12 cm
♂ Photo: A.Spreinat

A79130-5 *Maylandia* spec. "Zebra Blue Reef"

Lake Malawi: Minos Reef (Mozambique), 12 cm
♂ Photo: Ad Konings

A79130-5 *Maylandia* spec. "Zebra Blue Reef"

Lake Malawi: Minos Reef (Mozambique), 12 cm
♀ Photo: Ad Konings

A79131-5 *Maylandia* spec. "Zebra Blue Reef"

Lake Malawi: Ukhungu Reef (Mozambique), 12 cm
♂ Photo: A.Spreinat

A79131-5 *Maylandia* spec. "Zebra Blue Reef"

Lake Malawi: Ukhungu Reef (Mozambique), 12 cm
♀ Photo: A.Spreinat

A79132-5 *Maylandia* spec. "Zebra Blue Reef"

Lake Malawi: Ngumbe Rock/Minga Bay (Mozambique), 12 cm

Photo: A.Spreinat

A79120-5 *Maylandia* spec. "Zebra Blue"

Lake Malawi: Maleri Isl., 11 cm

Photo: A.Spreinat

A79120-5 *Maylandia* spec. "Zebra Blue"

Lake Malawi: Maleri Isl., 11 cm

Photo: Ad Konings

A79120-4 *Maylandia* spec. "Zebra Blue"

Lake Malawi: Maleri Isl., 11 cm

Photo: Ad Konings

A79150-5 *Maylandia* spec. "Zebra Chilumba"

Lake Malawi: Mara Rocks, 12 cm

Photo: Ad Konings

A79101-5 *Maylandia zebra* (Boulenger, 1899)

Lake Malawi: Cobue (Mozambique), 10-13 cm

Photo: Ad Konings

A79101-5 *Maylandia zebra* (Boulenger, 1899)
OB-Morph
Lake Malawi: Cobue (Mozambique), 10-13 cm

Photo: Ad Konings

A79101-5 *Maylandia zebra* (Boulenger, 1899)
OB-Morph
Lake Malawi: Cobue (Mozambique), 10-13 cm

Photo: Ad Konings

A79101-5 *Maylandia zebra* (Boulenger, 1899)

Lake Malawi: Cobue (Mozambique): north of Kanjindo, 10-13 cm
♂ Photo: A.Spreinat

A79055-5 *Maylandia zebra* (Boulenger, 1899)

Lake Malawi: Hai Reef (Tanzania), 10-13 cm
♂ Photo: A.Spreinat

A79055-5 *Maylandia zebra* (Boulenger, 1899)

Lake Malawi: Hai Reef (Tanzania), 10-13 cm
♀ Photo: A.Spreinat

A79055-5 *Maylandia zebra* (Boulenger, 1899)
O-Morph
Lake Malawi: Hai Reef (Tanzania), 10-13 cm
♀ Photo: A.Spreinat

A79076-5 *Maylandia zebra* (Boulenger, 1899)

Lake Malawi: Pombo Rocks (Tanzania), 10-13 cm
♂ Photo: A.Spreinat

A79076-5 *Maylandia zebra* (Boulenger, 1899)

Lake Malawi: Pombo Rocks (Tanzania), 10-13 cm
♂ Photo: A.Spreinat

A79102-5 *Maylandia zebra* (Boulenger, 1899)

Lake Malawi: Jalo, 10-13 cm
♂ Photo: Ad Konings

A79102-5 *Maylandia zebra* (Boulenger, 1899)

Lake Malawi: Jalo, 10-13 cm
♀ Photo: Ad Konings

A79069-5 *Maylandia zebra* (BOULENGER, 1899)

Lake Malawi: Ndumbi Reef (Tanzania), 10-13 cm

♂ Photo: A.Spreinat

A79306-5 *Maylandia zebra* (BOULENGER, 1899)

Lake Malawi: Undu Reef (Tanzania), 10-13 cm

♂ Photo: Ad Konings

A79306-5 *Maylandia zebra* (BOULENGER, 1899)

Lake Malawi: Undu Reef (Tanzania), 10-13 cm

♀ Photo: Ad Konings

A79306-5 *Maylandia zebra* (BOULENGER, 1899)
OB-Morph
Lake Malawi: Undu Reef (Tanzania), 10-13 cm

♀ Photo: Ad Konings

A79082-5 *Maylandia zebra* (BOULENGER, 1899)

Lake Malawi: Undu Point (Tanzania), 10-13 cm

♂ Photo: Ad Konings

A79082-5 *Maylandia zebra* (BOULENGER, 1899)

Lake Malawi: Undu Point (Tanzania), 10-13 cm

♂ Photo: A.Spreinat

A79307-5 *Maylandia zebra* (BOULENGER, 1899)

Lake Malawi: Pombo Reef (Tanzania), 10-13 cm

♂ Photo: Ad Konings

A79062-5 *Maylandia zebra* (BOULENGER, 1899)

Lake Malawi: Lundu (Tanzania), 10-13 cm

♂ Photo: Ad Konings

A79081-5 *Maylandia zebra* (Boulenger, 1899)

Lake Malawi: Tumbi Rocks (Tanzania), 10-13 cm

Photo: A.Spreinat

A79077-5 *Maylandia zebra* (Boulenger, 1899)

Lake Malawi: Puulu Isl. (Tanzania), 10-13 cm

Photo: E.Schraml

A79077-5 *Maylandia zebra* (Boulenger, 1899)
O + B-Morph
Lake Malawi: Puulu Isl. (Tanzania), 10-13 cm

Photo: A.Spreinat

A79077-5 *Maylandia zebra* (Boulenger, 1899)
OB-Morph
Lake Malawi: Puulu Isl. (Tanzania), 10-13 cm

Photo: E.Schraml

A79308-5 *Maylandia zebra* (Boulenger, 1899)

Lake Malawi: Puulu Isl. (Tanzania), 10-13 cm

Photo: A.Spreinat

A79061-5 *Maylandia zebra* (Boulenger, 1899)

Lake Malawi: Lundo Isl. (Tanzania), 10-13 cm

Photo: A.Spreinat

A79061-5 *Maylandia zebra* (Boulenger, 1899)
O-Morph
Lake Malawi: Lundo Isl. (Tanzania), 10-13 cm

Photo: A.Spreinat

A79061-5 *Maylandia zebra* (Boulenger, 1899)
OB-Morph
Lake Malawi: Lundo Isl. (Tanzania), 10-13 cm

Photo: A.Spreinat

A79060-5 *Maylandia zebra* (BOULENGER, 1899)

Lake Malawi: Mbamba Bay (Tanzania): Luhuchi Rocks, 10-13 cm

Photo: A.Spreinat

A79079-5 *Maylandia zebra* (BOULENGER, 1899)

Lake Malawi: Tumbi Point (Mozambique), 10-13 cm

Photo: A.Spreinat

A79309-5 *Maylandia zebra* (BOULENGER, 1899)

Lake Malawi: Cobue (Mozambique): Mara Point, 10-13 cm

Photo: A.Spreinat

A79309-5 *Maylandia zebra* (BOULENGER, 1899)

Lake Malawi: Cobue (Mozambique): Mara Point, 10-13 cm

Photo: Ad Konings

A79058-5 *Maylandia zebra* (BOULENGER, 1899)

Lake Malawi: Likoma: Membe Point, 10-13 cm

Photo: A.Spreinat

A79058-5 *Maylandia zebra* (BOULENGER, 1899)

Lake Malawi: Likoma, 10-13 cm

Photo: A.Spreinat

A79058-5 *Maylandia zebra* (BOULENGER, 1899)
O-Morph
Lake Malawi: Likoma, 10-13 cm

Photo: Ad Konings

A79058-5 *Maylandia zebra* (BOULENGER, 1899)
OB-Morph
Lake Malawi: Likoma, 10-13 cm

Photo: Ad Konings

A79058-5 *Maylandia zebra* (Boulenger, 1899)
OB-Morph
Lake Malawi: Likoma: Mbuzi Isl., 10-13 cm
Photo: A.Spreinat

A79078-5 *Maylandia zebra* (Boulenger, 1899)

Lake Malawi: Taiwan Reef, 10-13 cm
Photo: Ad Konings

A79311-5 *Maylandia zebra* (Boulenger, 1899) leg. Konings
OB-Morph, = *M. estherae* leg. Spreinat
Lake Malawi: Metangula (Mozambique), 10-13 cm
Photo: Ad Konings

A79311-5 *Maylandia zebra* (Boulenger, 1899) leg. Konings
= *M. estherae* leg. Spreinat
Lake Malawi: Metangula (Mozambique), 10-13 cm
Photo: Ad Konings

A79320-5 *Maylandia zebra* (Boulenger, 1899)
OB-Morph
Lake Malawi: Gome, 10-13 cm
Photo: Ad Konings

A79320-5 *Maylandia zebra* (Boulenger, 1899) leg. Konings
OB-Morph, = *M. estherae* (?) leg. Spreinat
Lake Malawi: Gome, 10-13 cm
Photo: A.Spreinat

A79080-5 *Maylandia zebra* (Boulenger, 1899)

Lake Malawi: Thumbi West Isl., 10-13 cm
Photo: A.Spreinat

A79316-5 *Maylandia zebra* (Boulenger, 1899)

Lake Malawi: Boadzulu Isl., 10-13 cm
Photo: Ad Konings

A79319-5 *Maylandia zebra* (Boulenger, 1899)

Lake Malawi: Namalenje Isl., 10-13 cm
♂ Photo: Ad Konings

A79317-5 *Maylandia zebra* (Boulenger, 1899)

Lake Malawi: Mazinzi Reef, 10-13 cm
♂ Photo: Ad Konings

A79051-5 *Maylandia zebra* (Boulenger, 1899)

Lake Malawi: Kande Isl.: Bandawe, 10-13 cm
♂ Photo: Ad Konings

A79072-5 *Maylandia zebra* (Boulenger, 1899)

Lake Malawi: Nkhata Bay: Chikale, 10-13 cm
♂ Photo: A.Spreinat

A79072-4 *Maylandia zebra* (Boulenger, 1899)

Lake Malawi: Nkhata Bay: Chindozwa, 10-13 cm
♀ Photo: A.Spreinat

A79072-5 *Maylandia zebra* (Boulenger, 1899)
O-Morph
Lake Malawi: Nkhata Bay, 10-13 cm
♀ Photo: A.Spreinat

A79072-5 *Maylandia zebra* (Boulenger, 1899)

Lake Malawi: Nkhata Bay?, 10-13 cm
♂ Photo: W.Staeck

A79050-5 *Maylandia zebra* (Boulenger, 1899)
OB-Morph
Lake Malawi, 10-13 cm
♂ Photo: W.Staeck

A79318-5 *Maylandia* cf. *melabranchion* (Stauffer, Bowers, Kellog & McKaye, 1997)

Lake Malawi: Zimbawe Rock, 13 cm

Photo: Ad Konings

A79315-5 *Maylandia zebra* (Boulenger, 1899)

Lake Malawi: Makanjila, 10-13 cm

Photo: A.Spreinat

A79315-5 *Maylandia zebra* (Boulenger, 1899)

Lake Malawi: Makanjila, 10-13 cm

Photo: Ad Konings

A79330-5 *Maylandia thapsinogen* (Stauffer, Bowers, Kellog & McKaye, 1997)

Lake Malawi: Eccles Reef, 13 cm

Photo: Ad Konings

A79335-5 *Maylandia sandaracinos* (Stauffer, Bowers, Kellog & McKaye, 1997)

Lake Malawi: Nkhudzi, 13 cm

Photo: Ad Konings

A79340-6 *Maylandia pyrsonotos* (Stauffer, Bowers, Kellog & McKaye, 1997)

Lake Malawi: Nakanthenga Isl., 13 cm

Photo: E.Schraml

A79340-5 *Maylandia pyrsonotos* (Stauffer, Bowers, Kellog & McKaye, 1997)

Lake Malawi: Nakanthenga Isl., 13 cm

Photo: A.Spreinat

A79341-5 *Maylandia pyrsonotos* (Stauffer, Bowers, Kellog & McKaye, 1997), O-Morph

Lake Malawi: Nakanthenga Isl., 13 cm

Photo: A.Spreinat

A79341-5 *Maylandia pyrsonotos* (Stauffer, Bowers, Kellog & McKaye, 1997) O-Morph
Lake Malawi: Nakanthenga Isl., 13 cm
♀ Photo: A.Spreinat

A79345-5 *Maylandia emmiltos* (Stauffer, Bowers, Kellog & McKaye, 1997)
Lake Malawi: Chilumba: Mpanga Rocks, 13 cm
♂ Photo: A.Spreinat

A79105-5 *Maylandia cyneusmarginatus* (Stauffer, Bowers, Kellog & McKaye, 1997)
Lake Malawi: Chia Lagoon, 10-11 cm
♂ Photo: Ad Konings

A79106-4 *Maylandia cyneusmarginatus* (Stauffer, Bowers, Kellog & McKaye, 1997)
Lake Malawi: Nkhomo Reef, 10-11 cm
♂ Photo: Ad Konings

A79195-5 *Maylandia* spec. "Zebra Gold Breast Orange Top"
Lake Malawi: Hongi Isl. (Tanzania), 11-13 cm
♂ Photo: A.Spreinat

A79192-5 *Maylandia* spec. "Zebra Gold Breast Orange Top"
Lake Malawi: Mbahwa Isl. (Tanzania), 11-13 cm
♂ Photo: A.Spreinat

A79191-5 *Maylandia* spec. "Zebra Gold Breast Orange Top"
Lake Malawi: Lundo Isl. (Tanzania), 11-13 cm
♂ Photo: A.Spreinat

A79191-6 *Maylandia* spec. "Zebra Gold Breast Orange Top"
Lake Malawi: Lundo Isl. (Tanzania), 11-13 cm
♂ Photo: A.Spreinat

A79190-5 *Maylandia* spec. "Zebra Gold Breast Orange Top"
O-Morph
Lake Malawi: Lundo Isl. (Tanzania), 11-13 cm
♂ Photo: A.Spreinat

A79190-5 *Maylandia* spec. "Zebra Gold Breast Orange Top"
OB-Morph
Lake Malawi: Lundo Isl. (Tanzania), 11-13 cm
♂ Photo: A.Spreinat

A79190-5 *Maylandia* spec. "Zebra Gold Breast Orange Top"
OB-Morph
Lake Malawi: Lundo Isl. (Tanzania), 11-13 cm
♀ Photo: A.Spreinat

A79191-5 *Maylandia* spec. "Zebra Gold Breast Orange Top"
Lake Malawi: Lundo Isl. (Tanzania), 11-13 cm
♀ Photo: A.Spreinat

A79193-5 *Maylandia* spec. "Zebra Gold Breast Orange Top"
Lake Malawi: Ndumbi Reef (Tanzania), 11-13 cm
♂ Photo: A.Spreinat

A79194-5 *Maylandia* spec. "Zebra Gold Breast Orange Top"
Lake Malawi: Pombo Rocks (Tanzania), 11-13 cm
♂ Photo: A.Spreinat

A79185-5 *Maylandia* spec. "Zebra Gold Breast Mbamba"
Lake Malawi: Mbamba Bay (Tanzania): Higga Reef, 11-13 cm
♂ Photo: A.Spreinat

A79185-5 *Maylandia* spec. "Zebra Gold Breast Mbamba"
Lake Malawi: Mbamba Bay (Tanzania): Higga Reef, 11-13 cm
♀ Photo: A.Spreinat

A79186-5 *Maylandia* spec. "Zebra Gold Breast Mbamba"

Lake Malawi: Mbamba Bay (Tanzania): Matalawe Reef, 11-13 cm

Photo: A.Spreinat

A79350-5 *Maylandia* spec. "Zebra Gold Breast Mozambique"

Lake Malawi: Metangula (Mozambique): Chuanga, 11-13 cm

Photo: A.Spreinat

A79351-5 *Maylandia* spec. "Zebra Gold Breast Mozambique"

Lake Malawi: N'nosi Reef (Mozambique), 11-13 cm

Photo: A.Spreinat

A79351-5 *Maylandia* spec. "Zebra Gold Breast Mozambique"

Lake Malawi: N'nosi Reef (Mozambique), 11-13 cm

Photo: A.Spreinat

A79352-5 *Maylandia* spec. "Zebra Gold Breast Mozambique"

Lake Malawi: Ukhungu Reef (Mozambique), 11-13 cm

Photo: A.Spreinat

A79353-5 *Maylandia* spec. "Zebra Gold Breast Mozambique"

Lake Malawi: Minga Bay (Mozambique): Ngumbe Rock, 11-13 cm

Photo: A.Spreinat

A79354-5 *Maylandia* spec. "Zebra Gold Breast Mozambique"

Lake Malawi: Chikala (Mozambique), 11-13 cm

Photo: A.Spreinat

A79355-5 *Maylandia* spec. "Zebra Gold Breast Mozambique"

Lake Malawi: Meponda (Mozambique): Nyamizimu, 11-13 cm

Photo: A.Spreinat

A79065-5 *Maylandia zebra* (Boulenger, 1899) leg. Konings
= *Maylandia* spec. "Zebra Gold Breast Mozambique" leg. Spreinat
Lake Malawi: Minos Reef (Mozambique), 10-13 cm
♂ Photo: Ad Konings

A79065-5 *Maylandia zebra* (Boulenger, 1899)leg. Konings
= *Maylandia* spec. "Zebra Gold Breast Mozambique" leg. Spreinat
Lake Malawi: Minos Reef (Mozambique), 10-13 cm
♀ Photo: Ad Konings

A79312-5 *Maylandia zebra* (Boulenger, 1899) leg. Konings
= *Maylandia* spec. "Zebra Gold Breast Mozambique" leg. Spreinat
Lake Malawi: Chiloelo (Mozambique), 10-13 cm
♂ Photo: Ad Konings

A79310-5 *Maylandia zebra* (Boulenger, 1899) leg. Konings
= *Maylandia* spec. "Zebra Gold Breast Mozambique" leg. Spreinat
Lake Malawi: Chuanga (Mozambique), 10-13 cm
♂ Photo: Ad Konings

A79314-5 *Maylandia zebra* (Boulenger, 1899) leg. Konings
= *Maylandia* spec. "Zebra Gold Breast Mozambique" leg. Spreinat
Lake Malawi: Thundu (Mozambique), 10-13 cm
♂ Photo: Ad Konings

A79073-5 *Maylandia zebra* (Boulenger, 1899) leg. Konings
= *Maylandia* spec. "Zebra Gold Breast Mozambique" leg. Spreinat
Lake Malawi: Nkhungu Reef (Mozambique), 10-13 cm
♂ Photo: Ad Konings

A79313-5 *Maylandia zebra* (Boulenger, 1899) leg. Konings
= *Maylandia* spec. "Zebra Gold Breast Mozambique" leg. Spreinat
Lake Malawi: Lumessi South (Mozambique), 10-13 cm
♂ Photo: Ad Konings

A79356-5 *Maylandia zebra* (Boulenger, 1899) leg. Konings
= *Maylandia* spec. "Zebra Gold Breast Mozambique" leg. Spreinat
Lake Malawi: Gome, 10-13 cm
♂ Photo: Ad Konings

A78384-5 *Maylandia fainzilberi* (STAECK, 1976)

Lake Malawi: Makonde (Tanzania), 11-13 cm

♂ Photo: Ad Konings

A78382-5 *Maylandia fainzilberi* (STAECK, 1976)

Lake Malawi: Ikombe (Tanzania), 11-13 cm

♂ Photo: A.Spreinat

A78381-5 *Maylandia fainzilberi* (STAECK, 1976)

Lake Malawi: Cove Mountain (Tanzania), 11-13 cm

♂ Photo: A.Spreinat

A78381-5 *Maylandia fainzilberi* (STAECK, 1976)

Lake Malawi: Cove Mountain (Tanzania), 11-13 cm

♀ Photo: A.Spreinat

A78380-5 *Maylandia fainzilberi* (STAECK, 1976)

Lake Malawi: (?) 11-13 cm

♂ Photo: W.Staeck

A78387-5 *Maylandia fainzilberi* (STAECK, 1976)
OB-Morph
Lake Malawi: (?) 11-13 cm

♂ Photo: E.Schraml

A78387-5 *Maylandia fainzilberi* (STAECK, 1976)
OB-Morph
Lake Malawi: Makonde (Tanzania), 11-13 cm

♂ Photo: Ad Konings

A78387-5 *Maylandia fainzilberi* (STAECK, 1976)
OB-Morph
Lake Malawi: (?) 11-13 cm

♂ Photo: A.Spreinat

A79025-5 *Maylandia xanstomachus* (Stauffer & Bolz, 1989)

Lake Malawi: Nakanthenga Isl., 12-13 cm

♂ Photo: A.Spreinat

A79025-5 *Maylandia xanstomachus* (Stauffer & Bolz, 1989)

Lake Malawi: Nakanthenga Isl., 12-13 cm

♂ Photo: A.Spreinat

A79025-5 *Maylandia xanstomachus* (Stauffer & Bolz, 1989)

Lake Malawi: Nakanthenga Isl., 12-13 cm

♂ Photo: A.Spreinat

A79026-5 *Maylandia xanstomachus* (Stauffer & Bolz, 1989)

Lake Malawi: Maleri Isl., 12-13 cm

♂ Photo: W.Deproost

A79027-5 *Maylandia xanstomachus* (Stauffer & Bolz, 1989)
OB-Morph
Lake Malawi: Maleri Isl., 12-13 cm

♂ Photo: A.Spreinat

A79027-5 *Maylandia xanstomachus* (Stauffer & Bolz, 1989)
OB-Morph
Lake Malawi: Maleri- /Nakanthenga Isl., 12-13 cm

♂ Photo: A.Spreinat

A79025-5 *Maylandia xanstomachus* (Stauffer & Bolz, 1989)
OB-Morph
Lake Malawi: Maleri-/Nakanthenga Isl., 12-13 cm

♂ ♀ Photo: A.Spreinat

A79027-5 *Maylandia xanstomachus* (Stauffer & Bolz, 1989)
OB-Morph
Lake Malawi: Maleri-/Nakanthenga Isl., 12-13 cm

♂ ♀ Photo: A.Spreinat

A79149-5 *Maylandia* spec. "Zebra Chilumba"

Lake Malawi: Mdoka, 13 cm

Photo: Ad Konings

A79147-5 *Maylandia* spec. "Zebra Chilumba"

Lake Malawi: Chilumba: Chitande, 13 cm

Photo: A.Spreinat

A79147-5 *Maylandia* spec. "Zebra Chilumba"

Lake Malawi: Chilumba: Chitande, 13 cm

Photo: Ad Konings

A79147-5 *Maylandia* spec. "Zebra Chilumba"

Lake Malawi: Chilumba: Chitande, 13 cm

Photo: E.Schraml

A79151-5 *Maylandia* spec. "Zebra Chilumba" leg. Konings
OB-Morph, *M. callainos*(?)
Lake Malawi: Chilumba: Chitande, 13 cm

Photo: Ad Konings

A79145-6 *Maylandia* spec. "Zebra Chilumba"

Lake Malawi: Chilumba, 13 cm

Photo: A.Spreinat

A79145-3 *Maylandia* spec. "Zebra Chilumba"

Lake Malawi: Chilumba, 13 cm

Photo: A.Spreinat

A79152-5 *Maylandia* spec. "Zebra Chilumba"

Lake Malawi: Chilumba: Mpanga Rocks, 13 cm

Photo: A.Spreinat

A79148-5 *Maylandia* spec. "Zebra Chilumba"

Lake Malawi: Katale, 13 cm

Photo: Ad Konings

A79153-5 *Maylandia* spec. "Zebra Chilumba"

Lake Malawi: Mbowe, 13 cm

Photo: Ad Konings

A79154-5 *Maylandia* spec. "Zebra Chilumba"

Lake Malawi: Usisya, 13 cm

Photo: A.Spreinat

A79145-6 *Maylandia* spec. "Zebra Chilumba"

Lake Malawi: Chilumba, 13 cm

Photo: W.Staeck

A79146-5 *Maylandia* spec. "Zebra Chilumba"

Lake Malawi: Chilumba, 13 cm

Photo: E.Schraml

A79230-5 *Maylandia* spec. "Zebra Mbowe"

Lake Malawi: Mbowe, 12 cm

Photo: Ad Konings

A79210-5 *Maylandia* spec. "Zebra Long Pelvic"

Lake Malawi: Chesese, 12 cm

Photo: Ad Konings

A79211-5 *Maylandia* spec. "Zebra Long Pelvic"

Lake Malawi: Chilumba: Chewere, 12 cm

Photo: A.Spreinat

A79286-5 *Maylandia* spec. "Zebra Yellow Belly"

Lake Malawi: Ndumbi Reef (Tanzania), 10-12 cm

Photo: A.Spreinat

A79286-5 *Maylandia* spec. "Zebra Yellow Belly"

Lake Malawi: Ndumbi Reef (Tanzania), 10-12 cm

Photo: A.Spreinat

A79287-5 *Maylandia* spec. "Zebra Yellow Belly"

Lake Malawi: Pombo Rocks (Tanzania), 10-12 cm

Photo: A.Spreinat

A79292-5 *Maylandia* spec. "Zebra Yellow Tail"

Lake Malawi: Magunga (Tanzania), 10-12 cm

Photo: A.Spreinat

A79291-5 *Maylandia* spec. "Zebra Yellow Tail"

Lake Malawi: Cove Mountain (Tanzania), 10-12 cm

Photo: A.Spreinat

A79293-5 *Maylandia* spec. "Zebra Yellow Tail"

Lake Malawi: Manda (Tanzania), 10-12 cm

Photo: A.Spreinat

A78541-5 *Maylandia* spec. "Mbamba Bay Kompakt"

Lake Malawi: Higga Reef, 10-11 cm

Photo: A.Spreinat

A78541-5 *Maylandia* spec. "Mbamba Bay Kompakt"

Lake Malawi: Higga Reef, 10-11 cm

Photo: A.Spreinat

A78540-5 *Maylandia* spec. "Mbamba Bay Kompakt"

Lake Malawi: Mbamba Bay, 10-11 cm

Photo: E.Schraml

A78542-5 *Maylandia* spec. "Mbamba Bay Kompakt"
OB-Morph
Lake Malawi: Mbamba Bay(Tanzania): Ngkuyo Isl., 10-11 cm

Photo: A.Spreinat

A79175-4 *Maylandia* spec. "Mustardi"
= *Maylandia* spec. "Zebra Gold"
Lake Malawi: Chirombo Point bis Charo, 13 cm

Photo: E.Schraml

A79175-4 *Maylandia* spec. "Mustardi"
= *Maylandia* spec. "Zebra Gold"
Lake Malawi: Chirombo Point bis Charo, 13 cm

Photo: E.Schraml

A79175-6 *Maylandia* spec. "Mustardi"
= *Maylandia* spec. "Zebra Gold"
Lake Malawi: Chirombo Point bis Charo, 13 cm

Photo: W.Staeck

A79180-5 *Maylandia* spec. "Mustardi"
= *Maylandia* spec. "Zebra Gold"
Lake Malawi: Nkhata Bay, 13 cm

Photo: Ad Konings

A79175-6 *Maylandia* spec. "Mustardi"
= *Maylandia* spec. "Zebra Gold"
Lake Malawi: Chirombo Point bis Charo, 13 cm

Photo: A.Spreinat

A79178-5 *Maylandia* spec. "Mustardi"
= *Maylandia* spec. "Zebra Gold"
Lake Malawi: Lion's Cove, 13 cm

Photo: A.Spreinat

A79178-5 *Maylandia* spec. "Mustardi"
= *Maylandia* spec. "Zebra Gold"
Lake Malawi: Lion's Cove, 13 cm
Photo: A.Spreinat

A79181-5 *Maylandia* spec. "Mustardi"
= *Maylandia* spec. "Zebra Gold", OB-Morph
Lake Malawi: Lion's Cove, 13 cm
Photo: Ad Konings

A79176-5 *Maylandia* spec. "Mustardi"
= *Maylandia* spec. "Zebra Gold"
Lake Malawi: Charo, 13 cm
Photo: Ad Konings

A79177-5 *Maylandia* spec. "Mustardi"
= *Maylandia* spec. "Zebra Gold", OB-Morph
Lake Malawi: Kakusa, 13 cm
Photo: Ad Konings

A79360-5 *Maylandia* spec. "Zebra Dwarf Gold"

Lake Malawi: Makanjila, 8-10 cm
Photo: A.Spreinat

A79360-5 *Maylandia* spec. "Zebra Dwarf Gold"

Lake Malawi: Makanjila, 8-10 cm
Photo: A.Spreinat

A79125-5 *Maylandia* spec. "Zebra Blue Gold"

Lake Malawi: Lupingu (Tanzania), 9-10 cm
Photo: A.Spreinat

A79125-5 *Maylandia* spec. "Zebra Blue Gold"

Lake Malawi: Lupingu (Tanzania), 9-10 cm
Photo: A.Spreinat

A79260-5 *Maylandia* spec. "Zebra Slim" leg. Spreinat
= *Maylandia* spec. "Kingsizei North" leg. Konings
Lake Malawi: Nkanda (Tanzania), 8-10 cm
♂ Photo: A.Spreinat

A79260-5 *Maylandia* spec. "Zebra Slim" leg. Spreinat
= *Maylandia* spec. "Kingsizei North" leg. Konings
Lake Malawi: Nkanda (Tanzania), 8-10 cm
♀ Photo: A.Spreinat

A79259-5 *Maylandia* spec. "Zebra Slim" leg. Spreinat
= *Maylandia* spec. "Kingsizei North" leg. Konings
Lake Malawi: Magunga (Tanzania), 8-10 cm
♂ Photo: A.Spreinat

A79259-5 *Maylandia* spec. "Zebra Slim" leg. Spreinat
= *Maylandia* spec. "Kingsizei North" leg. Konings
Lake Malawi: Magunga (Tanzania), 8-10 cm
♀ Photo: A.Spreinat

A79257-5 *Maylandia* spec. "Zebra Slim" leg. Spreinat
= *Maylandia* spec. "Kingsizei North" leg. Konings
Lake Malawi: Kirondo (Tanzania), 8-10 cm
♂ Photo: A.Spreinat

A79256-5 *Maylandia* spec. "Kingsizei North" leg. Konings
= *Maylandia spec.* "Zebra Slim" leg. Spreinat
Lake Malawi: Ikombe (Tanzania), 8-10 cm
♂ Photo: Ad Konings

A79258-5 *Maylandia* spec. "Kingsizei North" leg. Konings
= *Maylandia spec.* "Zebra Slim" leg. Spreinat
Lake Malawi: Lumbila (Tanzania), 8-10 cm
♀ Photo: Ad Konings

A79365-6 *Maylandia* spec. X *Labeotropheus fuelleborni*
Kreuzung / Cross-breed
Lake Malawi, 13 cm
♂ Photo: E.Schraml

A79166-5 *Maylandia* spec. "Zebra Dwarf Tansania"

Lake Malawi: Lupingu (Tanzania), 6-8 cm

Photo: A.Spreinat

A79168-5 *Maylandia* spec. "Zebra Dwarf Tansania"

Lake Malawi: Puulu (Tanzania), 6-8 cm

Photo: A.Spreinat

A79266-5 *Maylandia* spec. "Zebra South"

Lake Malawi: Hai Reef (Tanzania), 12-14 cm

Photo: A.Spreinat

A79267-5 *Maylandia* spec. "Zebra South"

Lake Malawi: Ponta Messuli (Tanzania), 12-14 cm

Photo: A.Spreinat

A79265-5 *Maylandia* spec. "Zebra South"

Lake Malawi: Undu Point (Tanzania), 12-14 cm

Photo: A.Spreinat

A79220-5 *Maylandia* spec. "Zebra Masinje"

Lake Malawi: N'tekete/Fort Maguire, 11-13 cm

Photo: A.Spreinat

A78040-5 *Maylandia aurora* (Burgess, 1976)

Lake Malawi, 11 cm

Photo: E.Schraml

A78040-5 *Maylandia aurora* (Burgess, 1976)

Lake Malawi, 11 cm

A78040-5 *Maylandia aurora* (Burgess, 1976)

Lake Malawi, 11 cm

♂ Photo: E.Schraml

A78040-5 *Maylandia aurora* (Burgess, 1976)

Lake Malawi, 11 cm

♀ Photo: A.Spreinat

A78043-5 *Maylandia aurora* (Burgess, 1976)

Lake Malawi: Cobue (Mozambique): Mara Point, 11 cm

♂ Photo: A.Spreinat

A78041-5 *Maylandia aurora* (Burgess, 1976)

Lake Malawi: Likoma: Mbamba Isl., 11 cm

♂ Photo: W.Deproost

A78042-5 *Maylandia aurora* (Burgess, 1976)

Lake Malawi: Tumbi Point (N) (Mozambique), 11 cm

♂ Photo: A.Spreinat

A78042-5 *Maylandia aurora* (Burgess, 1976)

Lake Malawi: Tumbi Point (Mozambique), 11 cm

♀ Photo: Ad Konings

A78057-5 *Maylandia* spec. "Aurora Uchesi"

Lake Malawi: Chicane (Mozambique): Uchesi Rocks, 11 cm

♂ Photo: A.Spreinat

A78057-5 *Maylandia* spec. "Aurora Uchesi"

Lake Malawi: Chicane (Mozambique): Uchesi Rocks, 11 cm

♀ Photo: A.Spreinat

A78055-5 *Maylandia* spec. "Aurora Lumbaulo"

Lake Malawi: Lumbaulo (Mozambique), 11 cm
♂ Photo: Ad Konings

A78055-5 *Maylandia* spec. "Aurora Lumbaulo"

Lake Malawi: Lumbaulo (Mozambique), 11 cm
♀ Photo: Ad Konings

A78060-5 *Maylandia* spec. "Aurora Yellow"

Lake Malawi: Lumessi (N) (Mozambique), 11 cm
♂ Photo: Ad Konings

A78060-5 *Maylandia* spec. "Aurora Yellow"

Lake Malawi: Lumessi (N) (Mozambique), 11 cm
♀ Photo: Ad Konings

A78050-5 *Maylandia* spec. "Aurora Blue"

Lake Malawi: Cobue (Mozambique), 11 cm
♂ Photo: Ad Konings

A78050-5 *Maylandia* spec. "Aurora Blue"

Lake Malawi: Cobue (Mozambique), 11 cm
♀ Photo: Ad Konings

A78051-5 *Maylandia* spec. "Aurora Blue"

Lake Malawi: Cobue (Mozambique): Kanjindo, 11 cm
♂ Photo: A.Spreinat

A78051-5 *Maylandia* spec. "Aurora Blue"

Lake Malawi: Cobue (Mozambique): Kanjindo, 11 cm
♀ Photo: A.Spreinat

A78052-5 *Maylandia* spec. "Aurora Blue"

Lake Malawi: Cobue (Mozambique): north of Kanjindo, 11 cm
♂ Photo: A.Spreinat

A78052-5 *Maylandia* spec. "Aurora Blue"

Lake Malawi: Cobue (Mozambique): north of Kanjindo, 11 cm
♀ Photo: A.Spreinat

A78053-5 *Maylandia* spec. "Aurora Blue"

Lake Malawi: Cobue (Mozambique): Namisi Rock, 11 cm
♂ Photo: A.Spreinat

A78053-5 *Maylandia* spec. "Aurora Blue"

Lake Malawi: Cobue (Mozambique): Namisi Rock, 11 cm
♀ Photo: A.Spreinat

A78046-5 *Maylandia* spec. "Aurora Black Tail"

Lake Malawi: Chiloelo (Mozambique): Ngulu, 11 cm
♂ Photo: A.Spreinat

A78046-5 *Maylandia* spec. "Aurora Black Tail"

Lake Malawi: Chiloelo (Mozambique): Ngulu, 11 cm
♀ Photo: A.Spreinat

A78045-4 *Maylandia* spec. "Aurora Black Tail"

Lake Malawi: Lumessi South (Mozambique): Ngulu, 11 cm
♂ Photo: Ad Konings

A78045-4 *Maylandia* spec. "Aurora Black Tail"

Lake Malawi: Lumessi South (Mozambique): Ngulu, 11 cm
♀ Photo: Ad Konings

A78080-4 *Maylandia* (?) / *Microchromis* (?) spec. "Black Dorsal Cobalt"

Lake Malawi: Likoma: Ndomo Bay, 12 cm

Photo: Ad Konings

A79110-5 *Maylandia* (?) / *Microchromis* (?) spec. "Zebra Bevous"

Lake Malawi: Chizumulu, 11 cm

Photo: Ad Konings

A79110-5 *Maylandia* (?) / *Microchromis* (?) spec. "Zebra Bevous"

Lake Malawi: Chizumulu: Membe Isl., 11 cm

Photo: A.Spreinat

A79110-5 *Maylandia* (?) / *Microchromis* (?) spec. "Zebra Bevous"

Lake Malawi: Chizumulu: Membe Isl., 11 cm

Photo: A.Spreinat

A79111-5 *Maylandia* (?) / *Microchromis* (?) spec. "Zebra Bevous"

Lake Malawi: Likoma: Membe Point, 11 cm

Photo: A.Spreinat

A79110-5 *Maylandia* (?) / *Microchromis* (?) spec. "Zebra Bevous"

Lake Malawi: Chizumulu: Membe Isl., 11 cm

Photo: A.Spreinat

A78700-4 *Pseudotropheus* spec. "Tiny"

Lake Malawi: Thumbi West, 8 cm

Photo: A.Spreinat

A78700-4 *Microchromis zebroides* Johnson, 1975

Lake Malawi: Likoma Isl.(?), 8 cm

Photo: D. Johnson

A79250-5 *Maylandia* spec. "Zebra Ruarwe"

Lake Malawi: Selewa, 12 cm

♂ Photo: Ad Konings

A79140-5 *Maylandia* spec. "Zebra Charo"

Lake Malawi: Charo, 12 cm

♂ Photo: Ad Konings

A78011-5 *Maylandia* spec. "Aggressive Bars"

Lake Malawi: Chiloelo (Mozambique), 12-14 cm

♂ Photo: Ad Konings

A78011-5 *Maylandia* spec. "Aggressive Bars"

Lake Malawi: Chiloelo (Mozambique), 12-14 cm

♀ Photo: Ad Konings

A78012-5 *Maylandia* spec. "Aggressive Bars"

Lake Malawi: Tumbi PTanzania (Mozambique), 12-14 cm

♂ Photo: Ad Konings

A78013-5 *Maylandia* spec. "Aggressive Bars"

Lake Malawi: Taiwan Reef, 12-14 cm

♂ Photo: Ad Konings

A78009-5 *Maylandia* spec. "Aggressive Bars"

Lake Malawi: Likoma: Ndumbi Rocks, 12-14 cm

♂ Photo: A.Spreinat

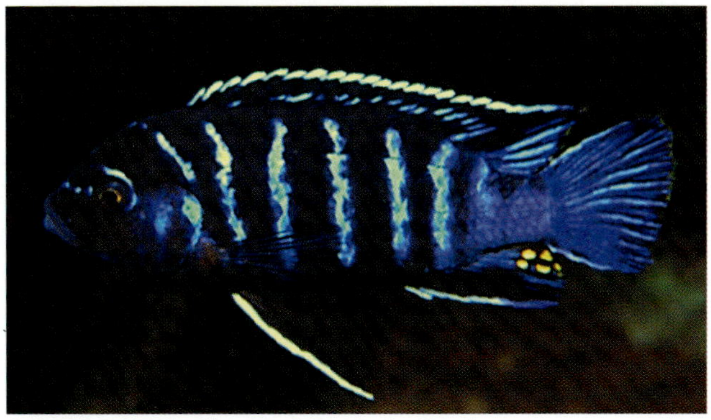

A78009-5 *Maylandia* spec. "Aggressive Bars"

Lake Malawi: Likoma: Ndumbi Rocks, 12-14 cm

♂ Photo: A.Spreinat

A78009-5 *Maylandia* spec. "Aggressive Bars"

Lake Malawi: Likoma: Ndumbi Rocks, 12-14 cm
♀ Photo: A.Spreinat

A78030-5 *Pseudotropheus* spec. "Aggressive Zebra"

Lake Malawi: Likoma: Ndomo Point, 12 cm
♂ Photo: A.Spreinat

A78030-5 *Pseudotropheus* spec. "Aggressive Zebra"

Lake Malawi: Likoma, 12 cm
♂ Photo: Ad Konings

A78030-5 *Pseudotropheus* spec. "Aggressive Zebra"

Lake Malawi: Likoma, 12 cm
♀ Photo: A.Konings

A78140-5 *Pseudotropheus* spec. "Chinyankwazi"

Lake Malawi: Chinyankwazi Isl., 12 cm
♂ Photo: A.Spreinat

A78140-5 *Pseudotropheus* spec. "Chinyankwazi"

Lake Malawi: Chinyankwazi Isl., 12 cm
♂ Photo: Ad Konings

A78102-5 *Maylandia phaeos* (STAUFFER, BOWERS, KELLOGG & MCKAYE, 1997)

Lake Malawi: Hai Reef (Tanzania), 12-13 cm
♂ Photo: A.Spreinat

A78102-5 *Maylandia phaeos* (STAUFFER, BOWERS, KELLOGG & MCKAYE, 1997)

Lake Malawi: Hai Reef (Tanzania), 12-13 cm
♀ Photo: A.Spreinat

A78105-5 *Maylandia phaeos* (Stauffer, Bowers, Kellogg & McKaye, 1997), before *Maylandia* spec. "Black Dorsal Tanzania"
Lake Malawi: Undu Point (Tanzania), 12-13 cm
♂ Photo: A.Spreinat

A78105-5 *Maylandia phaeos* (Stauffer, Bowers, Kellogg & McKaye, 1997), before *Maylandia* spec. "Black Dorsal Tanzania"
Lake Malawi: Undu Point (Tanzania), 12-13 cm
♀ Photo: A.Spreinat

A78103-5 *Maylandia phaeos* (Stauffer, Bowers, Kellogg & McKaye, 1997), before *Maylandia* spec. "Black Dorsal Tanzania"
Lake Malawi: Lundo Isl. (Tanzania), 12-13 cm
♂ Photo: A.Spreinat

A78103-5 *Maylandia phaeos* (Stauffer, Bowers, Kellogg & McKaye, 1997), before *Maylandia* spec. "Black Dorsal Tanzania"
Lake Malawi: Lundo Isl. (Tanzania), 12-13 cm
♀ Photo: A.Spreinat

A78098-5 *Maylandia phaeos* (Stauffer, Bowers, Kellogg & McKaye, 1997), before *Maylandia* spec. "Black Dorsal Shauri"
Lake Malawi: Cobue (Mozambique), 12-14 cm
♂ Photo: Ad Konings

A78098-5 *Maylandia phaeos* (Stauffer, Bowers, Kellogg & McKaye, 1997), before *Maylandia* spec. "Black Dorsal Shauri"
Lake Malawi: Cobue (Mozambique), 12-14 cm
♀ Photo: Ad Konings

A78099-5 *Maylandia phaeos* (Stauffer, Bowers, Kellogg & McKaye, 1997), before *Maylandia* spec. "Black Dorsal Shauri"
Lake Malawi: Londo (Mozambique), 12-14 cm
♂ Photo: Ad Konings

A78099-5 *Maylandia phaeos* (Stauffer, Bowers, Kellogg & McKaye, 1997), before *Maylandia* spec. "Black Dorsal Shauri"
Lake Malawi: Londo (Mozambique), 12-14 cm
♀ Photo: Ad Konings

A78088-5 *Maylandia phaeos* (Stauffer, Bowers, Kellogg & McKaye, 1997), before: *Maylandia* spec. "Black Dorsal Londo"
Lake Malawi: Londo/Ponta Messuli (Mozambique), 12 cm
♂ Photo: A.Spreinat

A78088-5 *Maylandia phaeos* (Stauffer, Bowers, Kellogg & McKaye, 1997), before: *Maylandia* spec. "Black Dorsal Londo"
Lake Malawi: Londo/Ponta Messuli (Mozambique), 12 cm
♀ Photo: A.Spreinat

A78070-5 *Maylandia* spec. "Black Dorsal"

Lake Malawi: Maleri Isl., 14 cm
♂ Photo: A.Spreinat

A78070-5 *Maylandia* spec. "Black Dorsal"

Lake Malawi: Maleri Isl., 14 cm
♀ Photo: A.Spreinat

A78070-5 *Maylandia* spec. "Black Dorsal"
also called *Pseudotropheus* spec. "Renati"
Lake Malawi,14cm
♂ Photo: E.Schraml

A78070-5 *Maylandia* spec. "Black Dorsal"
also called *Pseudotropheus* spec. "Renati"
Lake Malawi,14 cm
♀ Photo: E.Schraml

A78070-5 *Maylandia* spec. "Black Dorsal"

Lake Malawi: Maleri Isl., 14 cm
♂ Photo: A.Spreinat

A78073-4 *Maylandia* spec. "Black Dorsal"
Ribbink's *Ps.* cf. *heteropictus*
Lake Malawi: Thumbi West Isl., 14 cm
♂ Photo: A.Spreinat

A78073-4 *Maylandia* spec. "Black Dorsal"
Ribbink's *Ps.* cf. *heteropictus*
Lake Malawi: Thumbi West Isl., 14 cm
♂ Photo: A.Spreinat

A78073-4 *Maylandia* spec. "Black Dorsal"
Ribbink's *Ps.* cf. *heteropictus*
Lake Malawi: Thumbi West Isl., 14 cm
♀ Photo: A.Spreinat

A78075-5 *Maylandia* spec. "Black Dorsal Chiloelo"

Lake Malawi: Chiloelo (Mozambique): Mbalala, 14 cm
♂ Photo: A.Spreinat

A78075-5 *Maylandia* spec. "Black Dorsal Chiloelo"

Lake Malawi: Chiloelo (Mozambique), 14 cm
♀ Photo: Ad Konings

A78075-5 *Maylandia* spec. "Black Dorsal Chiloelo"

Lake Malawi: Meponda (Mozambique): Nyamizimu, 14 cm
♂ Photo: A.Spreinat

A78090-5 *Maylandia* spec. "Black Dorsal Nkhungu"

Lake Malawi: Minga Bay, Ngumbe Rock, 14 cm
♂ Photo: A.Spreinat

A78090-5 *Maylandia* spec. "Black Dorsal Nkhungu"

Lake Malawi: Nkhungu Point (Mozambique), 14 cm
♂ Photo: Ad Konings

A78090-5 *Maylandia* spec. "Black Dorsal Nkhungu"

Lake Malawi: Minga Bay, Ngumbe Rock, 14 cm
♀ Photo: A.Spreinat

A78085-5 *Maylandia* spec. "Black Dorsal Heteropictus"

Lake Malawi: Thumbi West, 12-14 cm

♂ Photo: Ad Konings

A78085-5 *Maylandia* spec. "Black Dorsal Heteropictus"

Lake Malawi: Thumbi West, 12-14 cm

♀ Photo: Ad Konings

A78110-5 *Maylandia* spec. "Black Dorsal Thundu"

Lake Malawi: Thundu (Mozambique), 12-14 cm

♂ Photo: Ad Konings

A78110-4 *Maylandia* spec. "Black Dorsal Thundu"

Lake Malawi: Thundu (Mozambique), 12-14 cm

♀ Photo: Ad Konings

A78095-5 *Maylandia* spec. "Black Dorsal Nkolongwe"

Lake Malawi, 12-14 cm

♂ Photo: W.Deproost

A78095-4 *Maylandia* spec. "Black Dorsal Nkolongwe"

Lake Malawi, 12-14 cm

♂ Photo: A.Spreinat

A78685-4 *Maylandia* spec. "Taiwan"

Lake Malawi: Taiwan Reef, 10 cm

♂ Photo: Ad Konings

A78685-4 *Maylandia* spec. "Taiwan"

Lake Malawi: Taiwan Reef, 10 cm

♀ Photo: Ad Konings

A78175-5 *Maylandia* spec. "Dolphin"

Lake Malawi: Puulu (Tanzania), 12-14 cm
♂ Photo: A.Spreinat

A78584-5 *Maylandia* spec. "Msobo"

Lake Malawi: Puulu (Tanzania), 8-10 cm
♂ Photo: A.Spreinat

A78584-4 *Maylandia* spec. "Msobo"

Lake Malawi: Puulu (Tanzania), 8-10 cm
♂ Photo: A.Spreinat

A78584-4 *Maylandia* spec. "Msobo"

Lake Malawi: Puulu (Tanzania), 8-10 cm
♀ Photo: A.Spreinat

A78577-5 *Maylandia* spec. "Msobo"

Lake Malawi: Lundo Isl. (Tanzania), 8-10 cm
♂ Photo: A.Spreinat

A78577-4 *Maylandia* spec. "Msobo"

Lake Malawi: Lundo Isl. (Tanzania), 8-10 cm
♂ Photo: A.Spreinat

A78577-4 *Maylandia* spec. "Msobo"

Lake Malawi: Lundo Isl. (Tanzania), 8-10 cm
♂ Photo: A.Spreinat

A78577-5 *Maylandia* spec. "Msobo"

Lake Malawi: Lundo Isl. (Tanzania), 8-10 cm
♀ Photo: A.Spreinat

A78577-2 *Maylandia* spec. "Msobo"

Lake Malawi: Lundo Isl. (Tanzania), 8-10 cm

Photo: A.Spreinat

A78579-3 *Maylandia* spec. "Msobo"

Lake Malawi: Magunga (Tanzania), 8-10 cm

♂

Photo: A.Spreinat

A78579-5 *Maylandia* spec. "Msobo"

Lake Malawi: Magunga (Tanzania), 8-10 cm

♂

Photo: A.Spreinat

A78579-5 *Maylandia* spec. "Msobo"

Lake Malawi: Magunga (Tanzania), 8-10 cm

♀

Photo: Ch.Wyrwich

A78578-5 *Maylandia* spec. "Msobo"

Lake Malawi: Lundu (Tanzania), 8-10 cm

♂

Photo: A.Spreinat

A78581-5 *Maylandia* spec. "Msobo"

Lake Malawi: Ndumbi Reef, 8-10 cm

♂

Photo: A.Spreinat

A78582-5 *Maylandia* spec. "Msobo"

Lake Malawi: Njambe (Tanzania), 8-10 cm

♂

Photo: A.Spreinat

A78590-4 *Maylandia* spec. "Msobo Heteropictus"

Lake Malawi: Puulu (Tanzania), 8-10 cm

♂

Photo: Ad Konings

A78595-5 *Pseudotropheus* spec. "Ndumbi Gold"
= M12
Lake Malawi: 12 cm
♂
Photo: A.Spreinat

A78595-5 *Pseudotropheus* spec. "Ndumbi Gold"
= M12
Lake Malawi: Likoma: Ndumbi Rocks,12 cm
♀
Photo: A.Spreinat

A78595-5 *Pseudotropheus* spec. "Ndumbi Gold"
= M12
Lake Malawi: Likoma: Mbako Point,12 cm
♂
Photo: A.Spreinat

A78595-4 *Pseudotropheus* spec. "Ndumbi Gold"
= M12
Lake Malawi: Likoma: Makulawe Point,12 cm
♂
Photo: A.Spreinat

A78425-5 *Pseudotropheus heteropictus* STAECK, 1980

Lake Malawi: Chisumulu Isl.,10 cm
♂
Photo: W.Staeck

A78425-5 *Pseudotropheus heteropictus* STAECK, 1980

Lake Malawi: Chisumulu Isl.,10 cm
♀
Photo: W.Staeck

A78425-5 *Pseudotropheus heteropictus* STAECK, 1980

Lake Malawi: Chisumulu Isl.,10 cm
♂
Photo: A.Spreinat

A78425-5 *Pseudotropheus heteropictus* STAECK, 1980

Lake Malawi: Chisumulu Isl.,10 cm
♀
Photo: A.Spreinat

A78442-5 *Pseudotropheus* spec. "Kingsizei"

Lake Malawi: Likoma: Maingano, 9 cm

♂ Photo: Ad Konings

A78442-5 *Pseudotropheus* spec. "Kingsizei"

Lake Malawi: Likoma: Maingano, 9 cm

♀ Photo: Ad Konings

A78441-5 *Pseudotropheus* spec. "Kingsizei"

Lake Malawi: Likoma: Mbuzi Isl., 9 cm

♂ Photo: A.Spreinat

A78440-5 *Pseudotropheus* spec. "Kingsizei"

Lake Malawi: Cobue (Mozambique): Kanjindo, 9 cm

♂ Photo: A.Spreinat

A78443-5 *Pseudotropheus* spec. "Kingsizei"

Lake Malawi: Ponta Messuli (Mozambique), 9 cm

♂ Photo: A.Spreinat

A78443-5 *Pseudotropheus* spec. "Kingsizei"

Lake Malawi: Ponta Messuli (Mozambique), 9 cm

♀ Photo: A.Spreinat

A78446-5 *Pseudotropheus* spec. "Kingsizei Londo"

Lake Malawi: Londo (Mozambique), 9 cm

♂ Photo: Ad Konings

A78446-4 *Pseudotropheus* spec. "Kingsizei Londo"

Lake Malawi: Londo (Mozambique), 9 cm

♀ Photo: Ad Konings

A78447-5 *Pseudotropheus* spec. "Kingsizei Londo"

Lake Malawi: Lumbaulo (Mozambique), 9 cm

♂ Photo: Ad Konings

A78447-5 *Pseudotropheus* spec. "Kingsizei Londo"

Lake Malawi: Lumbaulo (Mozambique), 9 cm

♀ Photo: Ad Konings

A78451-5 *Pseudotropheus* spec."Kingsizei Lupingu"

Lake Malawi: Lupingu (Tanzania), 9 cm

♂ Photo: Ad Konings

A78452-5 *Pseudotropheus* spec."Kingsizei Lupingu" leg. Konings
= *Maylandia* spec. "Zebra Blue Gold" leg. Spreinat
Lake Malawi: Lupingu (Tanzania), 9 cm

♂ Photo: Ad Konings

A78453-5 *Pseudotropheus* spec. "Kingsizei Lupingu" leg. Konings
= *Maylandia* spec. "Zebra Slim" leg. Spreinat
Lake Malawi: Ngwazi (Tanzania), 9 cm

♂ Photo: Ad Konings

A78453-5 *Pseudotropheus* spec. "Kingsizei Lupingu" leg. Konings
= *Maylandia* spec. "Zebra Slim" leg. Spreinat
Lake Malawi: Ngwazi (Tanzania), 9 cm

♀ Photo: Ad Konings

A78455-4 *Pseudotropheus* spec. "Kingsizei Tumbi"

Lake Malawi: Tumbi Point (Mozambique), 9 cm

♂ Photo: Ad Konings

A78455-3 *Pseudotropheus* spec. "Kingsizei Tumbi"

Lake Malawi: Tumbi Point (Mozambique), 9 cm

♀ Photo: Ad Konings

A78170-5 *Pseudotropheus demasoni* Konings, 1994

Lake Malawi: Pombo Reef (Tanzania), 6-8 cm
♂ Photo: M.K.Meyer

A78170-5 *Pseudotropheus demasoni* Konings, 1994
Pair
Lake Malawi: Pombo Reef (Tanzania), 6-8 cm
♂ ♀ Photo: A.Spreinat

A78170-4 *Pseudotropheus demasoni* Konings, 1994

Lake Malawi: Pombo Reef (Tanzania), 6-8 cm
♂ Photo: W.Deproost

A78170-5 *Pseudotropheus demasoni* Konings, 1994
Two males chasing each other
Lake Malawi: Pombo Reef (Tanzania), 6-8 cm
♂ ♂ Photo: A.Spreinat

A78670-5 *Pseudotropheus saulosi* Konings, 1990

Lake Malawi: Chisumulu: Taiwan Reef, 9 cm
♂ Photo: E.Schraml

A78670-5 *Pseudotropheus saulosi* Konings, 1990

Lake Malawi: Chisumulu: Taiwan Reef, 9 cm
♀ Photo: A.Spreinat

A78670-3 *Pseudotropheus saulosi* Konings, 1990

Lake Malawi: Chisumulu: Taiwan Reef, 9 cm
♂ Photo: F.Teigler/A.C.S.

A78670-3 *Pseudotropheus saulosi* Konings, 1990

Lake Malawi: Chisumulu: Taiwan Reef, 9 cm
♂ ♀ Photo: F.Teigler/A.C.S.

A78181-4 *Pseudotropheus* spec. "Dumpy"

Lake Malawi: Maleri Isl., 8 cm

Photo: A.Spreinat

A78182-4 *Pseudotropheus* spec. "Dumpy"

Lake Malawi: Nakatenga Isl., 8 cm

Photo: Ad Konings

A78183-4 *Pseudotropheus* spec. "Dumpy"

Lake Malawi: Ntekete, 8 cm

Photo: Ad Konings

A78545-5 *Pseudotropheus* spec. "Membe Deep"

Lake Malawi: Likoma: Membe Point, 10 cm

Photo: A.Spreinat

A78545-5 *Pseudotropheus* spec. "Membe Deep"

Lake Malawi: Likoma: Membe Point, 10 cm

Photo: Ad Konings

A78545-5 *Pseudotropheus* spec. "Membe Deep"

Lake Malawi: Likoma: Membe Point, 10 cm

Photo: Ad Konings

A78640-5 *Pseudotropheus* spec. "Polit"

Lake Malawi: Lion's Cove, 10 cm

Photo: A.Spreinat

A78645-4 *Pseudotropheus* spec. "Polit Tumbi"

Lake Malawi: Tumbi Point (Mozambique), 10 cm

Photo: Ad Konings

A78660-5 *Pseudotropheus* spec. "Red Top Ndumbi"

Lake Malawi: Ndumbi Reef (Tanzania), 7-8 cm

Photo: J.Steves

A78605-5 *Pseudotropheus* spec. "Orange Cap"

Lake Malawi: Pombo Reef (Tanzania), 8 cm

Photo: A.Spreinat

A78605-5 *Pseudotropheus* spec. "Orange Cap"

Lake Malawi: Undu Point (Tanzania), 8 cm

Photo: A.Spreinat

A78616-4 *Pseudotropheus perspicax* (TREWAVAS, 1935)

Lake Malawi: Mbowe Isl., 8-10 cm

Photo: Ad Konings

A78618-5 *Pseudotropheus perspicax* (TREWAVAS, 1935)

Lake Malawi: Mara Rocks, 8-10 cm

Photo: Ad Konings

A78615-5 *Pseudotropheus perspicax* (TREWAVAS, 1935)

Lake Malawi: Katale, 8-10 cm

Photo: Ad Konings

A78611-5 *Pseudotropheus perspicax* (TREWAVAS, 1935)

Lake Malawi: Chirwa Isl., 8-10 cm ♂

Photo: Ad Konings

A78611-5 *Pseudotropheus perspicax* (TREWAVAS, 1935)

Lake Malawi: Chirwa Isl., 8-10 cm ♀

Photo: Ad Konings

A78613-5 *Pseudotropheus perspicax* (Trewavas, 1935)
Lake Malawi: Hora Mhango, 8-10 cm
Photo: Ad Konings

A78612-5 *Pseudotropheus perspicax* (Trewavas, 1935)
Lake Malawi: Chitande, 8-10 cm
Photo: Ad Konings

A78614-5 *Pseudotropheus perspicax* (Trewavas, 1935)
Lake Malawi: Kakusa, 8-10 cm
Photo: Ad Konings

A78619-5 *Pseudotropheus perspicax* (Trewavas, 1935)
Lake Malawi: Chesese, 8-10 cm
Photo: Ad Konings

A78617-5 *Pseudotropheus perspicax* (Trewavas, 1935)
Lake Malawi: Mphanga Rocks, 8-10 cm
Photo: Ad Konings

A78624-5 *Pseudotropheus* spec. "Perspicax Tanzania" leg. Konings
= *Pseudotropheus* spec. "Variable Tanzania" leg. Spreinat
Lake Malawi: Tumbi Point (Tanzania), 8-10 cm
Photo: Ad Konings

A78626-5 *Pseudotropheus* spec. "Perspicax Tanzania" leg. Konings
= *Pseudotropheus* spec. "Variable Tanzania" leg. Spreinat
Lake Malawi: Njambe (Tanzania), 8-10 cm
Photo: Ad Konings

A78623-5 *Pseudotropheus* spec. "Perspicax Tanzania" leg. Konings
= *Pseudotropheus* spec. "Variable Tanzania" leg. Spreinat
Lake Malawi: Puulu (Tanzania), 8-10 cm
Photo: Ad Konings

A78622-5 *Pseudotropheus* spec. "Perspicax Tanzania" leg. KONINGS
= *Pseudotropheus* spec. "Variable Tanzania" leg. SPREINAT
Lake Malawi: Mbamba Bay (Tanzania), 7-9 cm
Photo: Ad Konings

A78622-5 *Pseudotropheus* spec. "Perspicax Tanzania" leg. KONINGS
= *Pseudotropheus* spec. "Variable Tanzania" leg. SPREINAT
Lake Malawi: Mbamba Bay (Tanzania), 7-9 cm
Photo: Ad Konings

A78621-5 *Pseudotropheus* spec. "Perspicax Tanzania" leg. KONINGS
= *Pseudotropheus* spec. "Variable Tanzania" leg. SPREINAT
Lake Malawi: Liuli (Tanzania), 7-9 cm
Photo: Ad Konings

A78625-5 *Pseudotropheus* spec. "Perspicax Tanzania" leg. KONINGS
= *Pseudotropheus* spec. "Variable Tanzania" leg. SPREINAT
Lake Malawi: Undu Reef (Tanzania), 7-9 cm
Photo: Ad Konings

A78997-5 *Pseudotropheus* spec. "Variable Tanzania" leg. SPREINAT
= *Pseudotropheus* spec. "Perspicax Tanzania" leg. KONINGS
Lake Malawi: Puulu (Tanzania), 7-9 cm
Photo: A.Spreinat

A78995-5 *Pseudotropheus* spec. "Variable Tanzania" leg. SPREINAT
= *Pseudotropheus* spec. "Perspicax Tanzania" leg. KONINGS
Lake Malawi: Tanzanian Coast, 7-9 cm
Photo: A.Spreinat

A78996-5 *Pseudotropheus* spec. "Variable Tanzania" leg. SPREINAT
= *Pseudotropheus* spec. "Perspicax Tanzania" leg. KONINGS
Lake Malawi: Lundo Isl. (Tanzania), 7-9 cm
Photo: A.Spreinat

A78998-5 *Pseudotropheus* spec. "Variable Tanzania" leg. SPREINAT
= *Pseudotropheus* spec. "Perspicax Tanzania" leg. KONINGS
Lake Malawi: Tumbi Rock (Tanzania), 7-9 cm
Photo: A.Spreinat

A78568-4 *Pseudotropheus* spec. "Minutus Mozambique"

Lake Malawi: Minos Reef (Mozambique), 7-9 cm

Photo: Ad Konings

A78568-4 *Pseudotropheus* spec. "Minutus Mozambique"

Lake Malawi: Minos Reef (Mozambique), 7-9 cm

Photo: Ad Konings

A78569-4 *Pseudotropheus* spec. "Minutus Mozambique"

Lake Malawi: Hai Reef (Tanzania), 7-9 cm

Photo: Ad Konings

A78569-4 *Pseudotropheus* spec. "Minutus Mozambique"

Lake Malawi: Hai Reef (Tanzania), 7-9 cm

Photo: Ad Konings

A78564-4 *Pseudotropheus* spec. "Minutus Mozambique"

Lake Malawi: Lupuchi (Mozambique), 7-9 cm

Photo: A.Spreinat

A78565-4 *Pseudotropheus* spec. "Minutus Mozambique"

Lake Malawi: Chikala (Mozambique), 7-9 cm

Photo: A.Spreinat

A78566-4 *Pseudotropheus* spec. "Minutus Mozambique"

Lake Malawi: Chilucha (Mozambique), 7-9 cm

Photo: Ad Konings

A78566-4 *Pseudotropheus* spec. "Minutus Mozambique"

Lake Malawi: Chilucha (Mozambique), 7-9 cm

Photo: Ad Konings

A78634-4 *Pseudotropheus* spec. "Minutus Tanzania" leg. Konings
= *Pseudotropheus* spec. "Plain" leg. Spreinat
Lake Malawi: Lupingu (Tanzania), 7-9 cm
Photo: Ad Konings

A78638-4 *Pseudotropheus* spec. "Minutus Tanzania" leg. Konings
= *Pseudotropheus* spec. "Plain" leg. Spreinat
Lake Malawi: Magunga (Tanzania), 7-9 cm
Photo: Ad Konings

A78635-4 *Pseudotropheus* spec. "Minutus Tanzania" leg. Konings
= *Pseudotropheus* spec. "Plain" leg. Spreinat
Lake Malawi: Manda (Tanzania), 7-9 cm
Photo: Ad Konings

A78635-4 *Pseudotropheus* spec. "Minutus Tanzania" leg. Konings
= *Pseudotropheus* spec. "Plain" leg. Spreinat
Lake Malawi: Manda (Tanzania), 7-9 cm
Photo: Ad Konings

A78636-4 *Pseudotropheus* spec. "Plain" leg. Spreinat
= *Pseudotropheus* spec. "Minutus Tanzania" leg. Konings
Lake Malawi: Nkanda (Tanzania), 7-9 cm
Photo: A.Spreinat

A78637-4 *Pseudotropheus* spec. "Minutus Tanzania" leg. Konings
= *Pseudotropheus* spec. "Plain" leg. Spreinat
Lake Malawi: Nkanda (Tanzania), 7-9 cm
Photo: Ad Konings

A78560-4 *Pseudotropheus minutus* Fryer, 1956

Lake Malawi: Nkhata Bay, 7-9 cm
Photo: A.Spreinat

A78560-4 *Pseudotropheus minutus* Fryer, 1956
Female?
Lake Malawi: Nkhata Bay, 7-9 cm
Photo: A.Spreinat

A78560-4 *Pseudotropheus minutus* Fryer, 1956

Lake Malawi: Nkhata Bay, 7-9 cm

Photo: W.Deproost

A78560-4 *Pseudotropheus minutus* Fryer, 1956

Lake Malawi: Nkhata Bay, 7-9 cm

Photo: A.Spreinat

A78985-4 *Pseudotropheus* spec. "Variable Eastern"

Lake Malawi: Masinje, 7-9 cm

Photo: Ad Konings

A78985-4 *Pseudotropheus* spec. "Variable Eastern"

Lake Malawi: Chiloelo (Mozambique), 7-9 cm

Photo: Ad Konings

A78990-5 *Pseudotropheus* spec. "Variable Kande"

Lake Malawi: Kande Isl., 7-9 cm

Photo: Ad Konings

A78993-4 *Pseudotropheus* spec. "Variable Jalo"

Lake Malawi: Jalo Reef, 7-9 cm

Photo: Ad Konings

A78646-5 *Pseudotropheus* spec. "Pombo Yellow Breast"

Lake Malawi: Pombo Reef (Tanzania), 7-9 cm

Photo: A.Spreinat

A78646-5 *Pseudotropheus* spec. "Pombo Yellow Breast"

Lake Malawi: Pombo Reef (Tanzania), 7-9 cm

Photo: A.Spreinat

A78653-5 *Pseudotropheus* cf. *purpuratus* JOHNSON, 1976

Lake Malawi: Meponda (Mozambique): Nyamizimu, 12 cm

Photo: A.Spreinat

A78650-5 *Pseudotropheus purpuratus* JOHNSON, 1976

Lake Malawi: Thumbi West Isl., 12 cm

Photo: Ad Konings

A78650-5 *Pseudotropheus purpuratus* JOHNSON, 1976

Lake Malawi: Maleri Isl., 12 cm

Photo: Ad Konings

A78650-5 *Pseudotropheus purpuratus* JOHNSON, 1976

Lake Malawi: Maleri Isl., 12 cm

Photo: Ad Konings

A78650-5 *Pseudotropheus purpuratus* JOHNSON, 1976

Lake Malawi: Gome, 12 cm

Photo: Ad Konings

A78655-5 *Pseudotropheus* spec. "Red Dorsal"

Lake Malawi: Mbenji Isl., 10 cm

Photo: A.Spreinat

A78655-5 *Pseudotropheus* spec. "Red Dorsal"

Lake Malawi: Mbenji Isl., 10 cm

Photo: A.Spreinat

A78655-5 *Pseudotropheus* spec. "Red Dorsal"

Lake Malawi: Mbenji Isl., 10 cm

Photo: A.Spreinat

A78010-5 *Pseudotropheus* spec. "Aggressive Brown"

Lake Malawi: Thumbi West Isl., 12 cm

Photo: Ad Konings

A78015-5 *Pseudotropheus* spec. "Aggressive Grey"

Lake Malawi: Likoma: Maingano, 12 cm

Photo: Ad Konings

A78015-5 *Pseudotropheus* spec. "Aggressive Grey"

Lake Malawi: Likoma: Ndomo Point, 12 cm

Photo: A.Spreinat

A78015-5 *Pseudotropheus* spec. "Aggressive Grey"

Lake Malawi: Likoma: Maingano, 12 cm

Photo: A.Spreinat

A78020-5 *Pseudotropheus* spec. "Aggressive Grey Head"

Lake Malawi: Nakanthenga, 11 cm

Photo: A.Spreinat

A78021-5 *Pseudotropheus* spec. "Aggressive Yellow Head"

Lake Malawi: Nakanthenga, 11 cm

Photo: A.Spreinat

A78502-5 *Pseudotropheus* spec. "Aggressive Puulu" leg. Spreinat
= *Pseudotropheus* spec. "Lucerna Blue Tanzania" leg. Konings
Lake Malawi: Puulu (N) (Tanzania), 12 cm

Photo: A.Spreinat

A78021-5 *Pseudotropheus* spec. "Aggressive Yellow Head"

Lake Malawi, 11 cm

Photo: E.Schraml

A78503-5 *Pseudotropheus* spec. "Lucerna Blue Tanzania" leg. Konings
= *Pseudotropheus* spec. "Aggressive Puulu" leg. Spreinat
Lake Malawi: Undu Point (Tanzania), 12 cm
♂ Photo: Ad Konings

A78503-4 *Pseudotropheus* spec. "Lucerna Blue Tanzania" leg. Konings
= *Pseudotropheus* spec. "Aggressive Puulu" leg. Spreinat
Lake Malawi: Undu Point (Tanzania), 12 cm
♀ Photo: Ad Konings

A78025-5 *Pseudotropheus* spec. "Aggressive Yellow Fin"

Lake Malawi: Chizumulu, 12 cm
♂ Photo: Ad Konings

A78025-5 *Pseudotropheus* spec. "Aggressive Yellow Fin"

Lake Malawi: Chizumulu, 12 cm
♂ Photo: A.Spreinat

A78390-5 *Pseudotropheus* spec. "Fin Spot"
= *Pseudotropheus* spec. "Lucerna Fin Spot"
Lake Malawi: Likoma Isl., 12 cm
♂ Photo: Ad Konings

A78390-5 *Pseudotropheus* spec. "Fin Spot"
= *Pseudotropheus* spec. "Lucerna Fin Spot"
Lake Malawi: Likoma Isl., 12 cm
♀ Photo: Ad Konings

A78497-5 *Pseudotropheus* spec. "Lucerna Blue Mozambique"

Lake Malawi: Metangula (Mozambique), 12 cm
♂ Photo: Ad Konings

A78497-5 *Pseudotropheus* spec. "Lucerna Blue Mozambique"

Lake Malawi: Metangula (Mozambique), 12 cm
♀ Photo: Ad Konings

A78715-5 *Pseudotropheus* spec. "Tropheops Aggressive"

Lake Malawi: Chewere, 12 cm

♂ Photo: Ad Konings

A78145-5 *Pseudotropheus* spec. "Cobalt"

Lake Malawi: Likoma: Ndomo Point, 12 cm

♂ Photo: A.Spreinat

A78145-5 *Pseudotropheus* spec. "Cobalt"
= *Pseudotropheus* spec. "Lucerna Blue Cobalt"
Lake Malawi: Likoma: Ndomo Point, 12 cm

♂ Photo: Ad Konings

A78495-5 *Pseudotropheus lucerna* Trewavas, 1935

Lake Malawi: Nkhata Bay, 12 cm

♂ Photo: Ad Konings

A78510-5 *Pseudotropheus* spec. "Lucerna Brown Chia"

Lake Malawi: near Chia Lagoon, 12 cm

♂ Photo: Ad Konings

A78510-4 *Pseudotropheus* spec. "Lucerna Brown Chia"

Lake Malawi: near Chia Lagoon, 12 cm

♀ Photo: Ad Konings

A79245-5 *Pseudotropheus* spec. "Patricki"

Lake Malawi: Mbenji Isl., 10 cm

♂ Photo: A.Spreinat

A79245-5 *Pseudotropheus* spec. "Patricki"

Lake Malawi, 10 cm

♂ Photo: A.Spreinat

A79245-5 *Pseudotropheus* spec. "Patricki"

Lake Malawi, 10 cm

♂

Photo: E.Schraml

A79245-5 *Pseudotropheus* spec. "Patricki"

Lake Malawi, 10 cm

♀

Photo: E.Schraml

A78004-5 *Pseudotropheus* spec. "Acei"

School/Schwarm

Lake Malawi: Ngara, 12 cm

Photo: Ad Konings

A78004-5 *Pseudotropheus* spec. "Acei"

Lake Malawi: Ngara, 12 cm

♂

Photo: Ad Konings

A78004-5 *Pseudotropheus* spec. "Acei"

Lake Malawi: Ngara, 12 cm

♀

Photo: Ad Konings

A78003-5 *Pseudotropheus* spec. "Acei"

Lake Malawi: Msuli, 12 cm

Photo: E.Schraml

A78003-5 *Pseudotropheus* spec. "Acei" mouthbreeding

Lake Malawi: Msuli, 12 cm

Photo: E.Schraml

A78001-5 *Pseudotropheus* spec. "Acei"

Lake Malawi: Chinteche, 12 cm

Photo: E.Schraml

A78005-5 *Pseudotropheus* spec. "Acei"

Lake Malawi: Senga Bay, 12 cm

Photo: A.Spreinat

A78006-5 *Pseudotropheus* spec. "Acei"

Lake Malawi: Ugaza, 12 cm

Photo: E.Schraml

A78435-5 *Pseudotropheus* spec. "Jacksoni"

Lake Malawi: Chisumulu, 13 cm

Photo: W.Staeck

A78435-5 *Pseudotropheus* spec. "Jacksoni"

Lake Malawi: Chisumulu, 13 cm

Photo: A.Spreinat

A78435-5 *Pseudotropheus* spec. "Jacksoni"

Lake Malawi: Chisumulu, 13 cm

Photo: A.Spreinat

A79004-5 *Pseudotropheus williamsi* (GÜNTHER, 1893)

Lake Malawi: Likoma Isl., 12-14 cm
♂ Photo: Ad Konings

A79004-5 *Pseudotropheus williamsi* (GÜNTHER, 1893)

Lake Malawi: Likoma Isl., 12-14 cm
♀ Photo: Ad Konings

A79003-5 *Pseudotropheus williamsi* (GÜNTHER, 1893)

Lake Malawi: Mbamba Bay (Tanzania): Chinula, 12-14 cm
♂ Photo: A.Spreinat

A79003-5 *Pseudotropheus williamsi* (GÜNTHER, 1893)

Lake Malawi: Mbamba Bay (Tanzania): Chinula, 12-14 cm
♀ Photo: A.Spreinat

A79001-5 *Pseudotropheus williamsi* (GÜNTHER, 1893)

Lake Malawi: Chizumulu Isl., 12-14 cm
♂ Photo: Ad Konings

A79022-5 *Pseudotropheus* spec. "Williamsi Tanzania"

Lake Malawi: Makonde (Tanzania), 12-14 cm
♂ Photo: Ad Konings

A79010-5 *Pseudotropheus* spec. "Williamsi Makanjila"

Lake Malawi: Tumbi Point (Mozambique), 12-14 cm
♂ Photo: Ad Konings

A79010-5 *Pseudotropheus* spec. "Williamsi Makanjila"

Lake Malawi: Lumessi North (Mozambique), 12-14 cm
♂ Photo: Ad Konings

A79015-5 *Pseudotropheus* spec. "Williamsi Maleri"

Lake Malawi: Namalenje Isl., 12-14 cm

Photo: d Konings

A79015-5 *Pseudotropheus* spec. "Williamsi Maleri"

Lake Malawi: Nakantenga Isl., 12-14 cm

Photo: Ad Konings

A79015-5 *Pseudotropheus* spec. "Williamsi Maleri"

Lake Malawi: Maleri Isl., 12-14 cm

Photo: E.Schraml

A79015-5 *Pseudotropheus* spec. "Williamsi Mbenji"
= *Pseudotropheus* spec. "Williamsi Maleri"
Lake Malawi: Mbenji Isl., 12-14 cm

Photo: A.Spreinat

A79020-5 *Pseudotropheus* spec. "Williamsi Nkudzi"

Lake Malawi: Boadzulu Isl., 12-14 cm

Photo: Ad Konings

A79020-5 *Pseudotropheus* spec. "Williamsi Nkudzi"

Lake Malawi: Boadzulu Isl., 12-14 cm

Photo: Ad Konings

A78969-4 *Pseudotropheus tursiops* Burgess & Axelrod, 1975
= *Pseudotropheus* spec. "Tursiops Chitande" leg. Konings
Lake Malawi: Ikombe (Tanzania), 12-14 cm

Photo: A.Spreinat

A78968-5 *Pseudotropheus tursiops* Burgess & Axelrod, 1975
= *Pseudotropheus* spec. "Tursiops Chitande" leg. Konings
Lake Malawi: Lupingu (Tanzania), 12-14 cm

Photo: A.Spreinat

A78972-5 *Pseudotropheus* spec. "Tursiops Chitande"

Lake Malawi: Makonde (Tanzania), 12-14 cm

Photo: Ad Konings

A78972-5 *Pseudotropheus* spec. "Tursiops Chitande"

Lake Malawi: Makonde (Tanzania), 12-14 cm

Photo: Ad Konings

A78973-5 *Pseudotropheus* spec. "Tursiops Chitande"

Lake Malawi: Chesese, 12-14 cm

Photo: Ad Konings

A78971-5 *Pseudotropheus* spec. "Tursiops Chitande"

Lake Malawi: Chitande Isl., 12-14 cm

Photo: Ad Konings

A78974-5 *Pseudotropheus* spec. "Tursiops Chitande"

Lake Malawi: Magunga (Tanzania), 12-14 cm

Photo: M.K.Meyer

A78974-5 *Pseudotropheus* spec. "Tursiops Chitande"

Lake Malawi: Magunga (Tanzania), 12-14 cm

Photo: Ad Konings

A78975-5 *Pseudotropheus* spec. "Tursiops Mbenji"

Lake Malawi: Mbenji Isl., 12-14 cm

Photo: Ad Konings

A78975-5 *Pseudotropheus* spec. "Tursiops Mbenji"

Lake Malawi: Mbenji Isl., 12-14 cm

Photo: A.Spreinat

A78966-5 *Pseudotropheus tursiops* Burgess & Axelrod, 1975

Lake Malawi: Chizumulu Isl., 12-14 cm
♂ Photo: Ad Konings

A78966-5 *Pseudotropheus tursiops* Burgess & Axelrod, 1975

Lake Malawi: Chizumulu Isl., 12-14 cm
♀ Photo: Ad Konings

A78150-5 *Pseudotropheus crabro* (Ribbink & Lewis, 1982)

Lake Malawi: Mbenji Isl., 15 cm
♂ Photo: E.Schraml

A78150-5 *Pseudotropheus crabro* (Ribbink & Lewis, 1982)

Lake Malawi: Mbenji Isl., 15 cm
♂ Photo: A.Spreinat

A78150-5 *Pseudotropheus crabro* (Ribbink & Lewis, 1982)

Lake Malawi: Mbenji Isl., 15 cm
♀ Photo: E.Schraml

A78150-5 *Pseudotropheus crabro* (*RIBBINK & LEWIS*, 1982)

Lake Malawi: N'tekete/Fort Maguire, 15 cm
♀ Photo: A.Spreinat

A78150-1 *Pseudotropheus crabro* (Ribbink & Lewis, 1982)

Lake Malawi: Mbenji Isl., 15 cm
Photo: E.Schraml

A78150-2 *Pseudotropheus crabro* (Ribbink & Lewis, 1982)

Lake Malawi: Meponda (Mozambique): Nyamizimu, 15 cm
Photo: A.Spreinat

A78675-5 *Pseudotropheus socolofi* Johnson, 1974

Lake Malawi: From Tumbi Point to Cobue (Mozambique), 12 cm

Photo: W.Deproost

A78675-5 *Pseudotropheus socolofi* Johnson, 1974

Lake Malawi: From Tumbi Point to Cobue (Mozambique), 12 cm

Photo: A.Spreinat

A78675-4 *Pseudotropheus socolofi* Johnson, 1974

Lake Malawi: Cobue (Mozambique): Namisi Rock, 12 cm

Photo: A.Spreinat

A78675-5 *Pseudotropheus socolofi* Johnson, 1974

Lake Malawi: From Tumbi Point to Cobue (Mozambique), 12 cm

Photo: E.Schraml

A78675-4 *Pseudotropheus socolofi* Johnson, 1974

Lake Malawi: Cobue (Mozambique): Kanjindo, 12 cm

Photo: A.Spreinat

A78120-4 *Pseudotropheus* spec. "Burower"

Lake Malawi: Nakantenga Isl., 10 cm

Photo: W.Deproost

A78120-5 *Pseudotropheus* spec. "Burower"

Lake Malawi: Maleri Isl., 10 cm

Photo: A.Spreinat

A78120-4 *Pseudotropheus* spec. "Burower"

Lake Malawi: Maleri Isl., 10 cm

Photo: A.Spreinat

A78161-5 *Pseudotropheus* spec. "Daktari Blue"
=? *Pseudotropheus* spec. "Daktari (Chiwindi)"
Lake Malawi: Mozambikan Coast, 10 cm
Photo: E.Schraml

A78161-5 *Pseudotropheus* spec. "Daktari Blue"
=? *Pseudotropheus* spec. "Daktari (Chiwindi)"
Lake Malawi: Mozambikan Coast, 10 cm
Photo: E.Schraml

A78162-5 *Pseudotropheus* spec. "Daktari"

Lake Malawi: Chiwindi (Mozambique), 10 cm
Photo: W.Deproost

A78162-5 *Pseudotropheus* spec. "Daktari"

Lake Malawi: Chiwindi (Mozambique), 10 cm
Photo: Ad Konings

A78165-5 *Pseudotropheus* spec. "Daktari Blue Gold"
=? *Pseudotropheus* spec. "Daktari (Chiwindi)"
Lake Malawi, 10 cm
Photo: E.Schraml

A78165-5 *Pseudotropheus* spec. "Daktari Blue Gold"
=? *Pseudotropheus* spec. *"Daktari* (Chiwindi)"
Lake Malawi, 10 cm
Photo: E.Schraml

A78163-5 *Pseudotropheus* spec. "Daktari"

Lake Malawi: Hai Reef (Tanzania), 10 cm
Photo: A.Spreinat

A78163-5 *Pseudotropheus* spec. "Daktari"

Lake Malawi: Hai Reef (Tanzania), 10 cm
Photo: A.Spreinat

A78160-5 *Pseudotropheus* spec. "Daktari"

Lake Malawi: Tanzanian Coast, 10 cm
♂ Photo: E.Schraml

A78164-5 *Pseudotropheus* spec. "Daktari"

Lake Malawi: Undu Reef (Tanzania), 10 cm
♂ Photo: Ad Konings

A78185-5 *Pseudotropheus* spec. "Dwarf Gold"

Lake Malawi: Ntekete, 9 cm
♂ Photo: Ad Konings

A78462-4 *Pseudotropheus* spec. "Lime"

Lake Malawi: Likoma: Membe Point, 9 cm
♂ Photo: Ad Konings

A78461-4 *Pseudotropheus* spec. "Lime"

Lake Malawi: Likoma: Mbamba, 9 cm
♂ Photo: Ad Konings

A78461-4 *Pseudotropheus* spec. "Lime"

Lake Malawi: Likoma: Mbamba, 9 cm
♀ Photo: Ad Konings

A78465-3 *Pseudotropheus* spec. "Lime Nkhomo"

Lake Malawi: Mbenji Isl., 10 cm
♂ Photo: Ad Konings

A78465-3 *Pseudotropheus* spec. "Lime Nkhomo"

Lake Malawi: Mbenji Isl., 10 cm
♀ Photo: Ad Konings

A78465-4 *Pseudotropheus* spec. "Lime Nkhomo"
spawning pair
Lake Malawi: Mbenji Isl., 10 cm
Photo: A.Spreinat

A78465-5 *Pseudotropheus* spec. "Lime Nkhomo"

Lake Malawi: Nkhomo Reef, 10 cm
Photo: Ad Konings

A78248-5 *Pseudotropheus* spec. "Elongatus Chinyamwezi"
=? *Pseudotropheus* spec. "Lime Nkhomo"
Lake Malawi: Chinyamwezi Isl., 10 cm
Photo: E.Schraml

A78400-5 *Pseudotropheus fuscus* Trewavas, 1935

Lake Malawi: Nkhata Bay: Chindozwa, 10 cm
Photo: A.Spreinat

A78400-5 *Pseudotropheus fuscus* Trewavas, 1935

Lake Malawi: Nkhata Bay, 10 cm
Photo: Ad Konings

A78405-5 *Pseudotropheus* spec. "Fuscus Dwarf"

Lake Malawi: Chuanga (Mozambique), 10 cm
Photo: Ad Konings

A78405-4 *Pseudotropheus* spec. "Fuscus Dwarf"

Lake Malawi: Londo (Mozambique), 10 cm
Photo: Ad Konings

A78405-3 *Pseudotropheus* spec. "Fuscus Dwarf"

Lake Malawi: Londo (Mozambique), 10 cm
Photo: Ad Konings

A78690-5 *Pseudotropheus* spec. "Thin Stripe"

Lake Malawi: Likoma: Membe Point, 8 cm

Photo: A.Spreinat

A78690-5 *Pseudotropheus* spec. "Thin Stripe"

Lake Malawi: Likoma: Mbako Point, 8 cm

Photo: A.Spreinat

A78690-5 *Pseudotropheus* spec. "Thin Stripe"

Lake Malawi: Likoma, 8 cm

Photo: Ad Konings

A78691-5 *Pseudotropheus* spec. "Thin Stripe"

Lake Malawi: Hai Reef (Tanzania), 8 cm

Photo: Ad Konings

A78115-5 *Pseudotropheus* spec. "Broad Bar" leg. Spreinat
= *Pseudotropheus* spec. "Crabro Blue" leg. Konings
Lake Malawi: Nkanda (Tanzania), 10-13 cm

Photo: A.Spreinat

A78115-5 *Pseudotropheus* spec. "Broad Bar" leg. Spreinat
= *Pseudotropheus* spec. "Crabro Blue" leg. Konings
Lake Malawi: Lupingu (Tanzania), 10-13 cm

Photo: A.Spreinat

A78035-5 *Pseudotropheus ater* Stauffer, 1988

Lake Malawi: Chinyamwezi Isl., 8 cm

Photo: A.Spreinat

A78035-3 *Pseudotropheus ater* Stauffer, 1988

Lake Malawi: Chinyamwezi Isl., 8 cm

Photo: A.Spreinat

A78155-5 *Pseudotropheus cyaneus* STAUFFER, 1988

Lake Malawi: Chinyamwezi Isl., 8-10 cm

♂

Photo: A.Spreinat

A78155-5 *Pseudotropheus cyaneus* STAUFFER, 1988

Lake Malawi: Chinyamwezi Isl., 8-10 cm

♀

Photo: A.Spreinat

A78155-5 *Pseudotropheus cyaneus* STAUFFER, 1988

Lake Malawi: Chinyamwezi Isl., 8-10 cm

♂

Photo: Ad Konings

A78155-4 *Pseudotropheus cyaneus* STAUFFER, 1988

Lake Malawi: Chinyamwezi Isl., 8-10 cm

♀

Photo: Ad Konings

A78571-4 *Pseudotropheus* spec. "Elongatus Bar"
=? *Pseudotropheus modestus* JOHNSON, 1974
Lake Malawi: Maleri Isl., 8 cm

♂

Photo: A.Spreinat

A78572-4 *Pseudotropheus modestus* JOHNSON, 1974

Lake Malawi: Chidunga Rocks, 8 cm

♂

Photo: Ad Konings

A78570-5 *Pseudotropheus modestus* JOHNSON, 1974

Lake Malawi: Nakantenga Isl., 8 cm

♂

Photo: Ad Konings

A78570-5 *Pseudotropheus modestus* JOHNSON, 1974
Male right top, female left
Lake Malawi: Maleri Isl., 8 cm

♀ ♂

Photo: D.S.Johnson

A78206-4 *Pseudotropheus elongatus* FRYER, 1956

Lake Malawi: Mbamba Bay (Tanzania): Mara Rocks, 12 cm

Photo: A.Spreinat

A78206-4 *Pseudotropheus elongatus* FRYER, 1956

Lake Malawi: Mbamba Bay (Tanzania): Mara Rocks, 12 cm

Photo: A.Spreinat

A78206-4 *Pseudotropheus elongatus* FRYER, 1956

Lake Malawi: Mbamba Bay (Tanzania): Ponton, 12 cm

Photo: A.Spreinat

A78293-4 *Pseudotropheus* spec. "Elongatus Mbako"

Lake Malawi: Likoma: Maingano, 10 cm

Photo: A.Spreinat

A78290-4 *Pseudotropheus* spec. "Elongatus Mbako"

Lake Malawi: Likoma: Mbako Point, 10 cm

Photo: W.Deproost

A78291-4 *Pseudotropheus* spec. "Elongatus Mbako"

Lake Malawi: Likoma: Makulawe, 10 cm

Photo: Ad Konings

A78292-4 *Pseudotropheus* spec. "Elongatus Mbako"

Lake Malawi: Likoma: Ndumbi Rocks, 10 cm

Photo: Ad Konings

A78273-5 *Pseudotropheus* spec. "Elongatus Hai"

Lake Malawi: Hai Reef (Tanzania), 10 cm

Photo: Ad Konings

A78305-5 *Pseudotropheus* spec. "Elongatus Metangula"

Lake Malawi: N'kolongwe (Mozambique), 10 cm

Photo: Ad Konings

A78305-4 *Pseudotropheus* spec. "Elongatus Metangula"

Lake Malawi: Lingwezi (Mozambique), 10 cm

Photo: A.Spreinat

A78346-5 *Pseudotropheus* spec. "Elongatus Reef"

Lake Malawi: Eccles Reef, 10 cm

Photo: A.Spreinat

A78347-5 *Pseudotropheus* spec. "Elongatus Reef"

Lake Malawi: Gome, 10 cm

Photo: Ad Konings

A78255-4 *Pseudotropheus* spec. "Elongatus Chisumulu"

Lake Malawi: Chizumulu Isl., 10 cm

Photo: A.Spreinat

A78255-4 *Pseudotropheus* spec. "Elongatus Chisumulu"

Lake Malawi: Chizumulu Isl., 10 cm

Photo: Ad Konings

A78320-5 *Pseudotropheus* spec. "Elongatus Namalenje"

Lake Malawi: Namalenje Isl., 10 cm

Photo: A. Spreinat

A78320-5 *Pseudotropheus* spec. "Elongatus Namalenje"

Lake Malawi: Namalenje Isl., 10 cm

Photo: Ad Konings

A78350-5 *Pseudotropheus* spec. "Elongatus Slab"

Lake Malawi: Makokola Reef, 10 cm

Photo: A.Spreinat

A78351-5 *Pseudotropheus* spec. "Elongatus Slab"

Lake Malawi: Monkey Bay: Tsano Rock, 10 cm

Photo: A.Spreinat

A78352-5 *Pseudotropheus* spec. "Elongatus Slab"

Lake Malawi: Maleri Isl., 10 cm

Photo: Ad Konings

A78353-5 *Pseudotropheus* spec. "Elongatus Slab"

Lake Malawi: Mumbo Isl., 10 cm

Photo: Ad Konings

A78355-5 *Pseudotropheus* spec. "Taiwan" leg. Konings
= *Pseudotropheus* spec. "Elongatus Deep Water" leg. Spreinat
Lake Malawi: Taiwan Reef, 10 cm

Photo: Ad Konings

A78355-5 *Pseudotropheus* spec. "Taiwan" leg. Konings
= *Pseudotropheus* spec. "Elongatus Deep Water" leg. Spreinat
Lake Malawi: Taiwan Reef, 10 cm

Photo: Ad Konings

A78263-5 *Pseudotropheus* spec. "Elongatus Deep Water" leg. Spreinat
= *Pseudotropheus* spec. "Taiwan" leg. Konings
Lake Malawi: Mbamba Bay (Tanzania): Matalawe Reef, 10-12 cm

Photo: A.Spreinat

A78263-5 *Pseudotropheus* spec. "Elongatus Deep Water" leg. Spreinat
= *Pseudotropheus* spec. "Taiwan" leg. Konings
Lake Malawi: Mbamba Bay (Tanzania), 10-12 cm

Photo: A.Spreinat

A78285-5 *Pseudotropheus* spec. "Elongatus Luhuchi"

Lake Malawi: Mbamba Bay (Tanzania): Luhuchi Rocks, 10 cm
♂ Photo: E.Schraml

A78285-4 *Pseudotropheus* spec. "Elongatus Luhuchi"

Lake Malawi: Mbamba Bay (Tanzania): Luhuchi Rocks, 10 cm
♀ Photo: A.Spreinat

A78285-5 *Pseudotropheus* spec. "Elongatus Luhuchi"
Yellow Morph
Lake Malawi: Mbamba Bay (Tanzania): Luhuchi Rocks, 10 cm
♀ Photo: A.Spreinat

A78892-3 *Pseudotropheus (Tropheops)* spec. "Tropheops Red Fin"

Lake Malawi: Hongi Isl. (Tanzania), 8-10 cm
♀ Photo: M.K.Meyer

A78340-5 *Pseudotropheus* spec. "Elongatus Ornatus"

Lake Malawi, 8-10 cm
♂ Photo: E.Schraml

A78340-4 *Pseudotropheus* spec. "Elongatus Ornatus"

Lake Malawi, 8-10 cm
♀ Photo: W.Deproost

A78340-5 *Pseudotropheus* spec. "Elongatus Ornatus"

Lake Malawi: Likoma: Ndumbi Rocks, 8-10 cm
♂ Photo: A.Spreinat

A78340-5 *Pseudotropheus* spec. "Elongatus Ornatus"

Lake Malawi: Likoma: Ndumbi Rocks, 8-10 cm
♀ Photo: A.Spreinat

A78340-5 *Pseudotropheus* spec. "Elongatus Ornatus"
= *Pseudotropheus* spec. "Ornatus"
Lake Malawi, 8-10 cm
♂ Photo: E.Schraml

A78340-5 *Pseudotropheus* spec. "Elongatus Ornatus"
= *Pseudotropheus* spec. "Ornatus"
Lake Malawi, 8-10 cm
♀ Photo: E.Schraml

A78340-5 *Pseudotropheus* spec. "Elongatus Ornatus"
= *Pseudotropheus* spec. "Ornatus"
Lake Malawi, 8-10 cm
♂ Photo: E.Schraml

A78340-5 *Pseudotropheus* spec. "Elongatus Ornatus"
= *Pseudotropheus* spec. "Ornatus"
Lake Malawi, 8-10 cm
♂ Photo: E.Schraml

A78310-4 *Pseudotropheus* spec. "Elongatus Mozambique Brown"

Lake Malawi: Lumbaulo (Mozambique), 8-10 cm
♂ Photo: Ad Konings

A78310-4 *Pseudotropheus* spec. "Elongatus Mozambique Brown"

Lake Malawi: Lumbaulo (Mozambique), 8-10 cm
♀ Photo: Ad Konings

A78300-5 *Pseudotropheus* spec. "Elongatus Mbenji Brown"

Lake Malawi: Mbenji Isl., 8-10 cm
♂ Photo: Ad Konings

A78300-5 *Pseudotropheus* spec. "Elongatus Mbenji Brown"

Lake Malawi: Mbenji Isl., 8-10 cm
♀ Photo: Ad Konings

A78335-4 *Pseudotropheus* spec. "Elongatus Nkhata Brown"

Lake Malawi: Nkhata Bay, 8-9 cm

Photo: Ad Konings

A78335-4 *Pseudotropheus* spec. "Elongatus Nkhata Brown"

Lake Malawi: Nkhata Bay: Chikale, 8-9 cm

Photo: A.Spreinat

A78325-4 *Pseudotropheus* spec. "Elongatus Ndumbi"

Lake Malawi: Likoma: Ndumbi Rocks, 8-9 cm

Photo: Ad Konings

A78280-4 *Pseudotropheus* spec. "Elongatus Likoma"

Lake Malawi: Likoma?, 8-10 cm

Photo: Ad Konings

A78330-4 *Pseudotropheus* spec. "Elongatus Nkhata Blue"

Lake Malawi: Kawanga, 8-10 cm

Photo: Ad Konings

A78491-4 *Pseudotropheus* spec. "Elongatus Spot"

Lake Malawi: Hongi Isl. (Tanzania), 10 cm

Photo: A.Spreinat

A78490-5 *Pseudotropheus* spec. "Elongatus Spot"

Lake Malawi: Puulu (Tanzania), 10 cm

Photo: A.Spreinat

A78490-5 *Pseudotropheus* spec. "Elongatus Spot"

Lake Malawi: Puulu (Tanzania), 10 cm

Photo: A.Spreinat

A78492-5 *Pseudotropheus* spec. "Elongatus Spot"

Lake Malawi: Lundo Isl. (Tanzania), 10 cm

Photo: A.Spreinat

A78488-4 *Pseudotropheus longior* SEEGERS, 1996

Lake Malawi: Mbamba Bay (Tanzania): Chinula, 10 cm

Photo: A.Spreinat

A78487-5 *Pseudotropheus longior* SEEGERS, 1996

Lake Malawi: Mbamba Bay (Tanzania): Ngkuyo Isl., 10 cm

Photo: A.Spreinat

A78487-4 *Pseudotropheus longior* SEEGERS, 1996

Lake Malawi: Mbamba Bay (Tanzania): Ngkuyo Isl., 10 cm

Photo: A.Spreinat

A78489-5 *Pseudotropheus longior* SEEGERS, 1996

Lake Malawi: Tumbi Point (Tanzania), 12 cm

Photo: Ad Konings

A78489-5 *Pseudotropheus longior* SEEGERS, 1996

Lake Malawi: Tumbi Point (Tanzania), 12 cm

Photo: Ad Konings

A78328-5 *Pseudotropheus* spec. "Elongatus Ngkuyo"

Lake Malawi: Mbamba Bay (Tanzania): Higga Reef,10 cm

Photo: A.Spreinat

A78328-5 *Pseudotropheus* spec. "Elongatus Ngkuyo"

Lake Malawi: Mbamba Bay (Tanzania): Ngkuyo Isl.,10 cm

Photo: A.Spreinat

A78210-5 *Pseudotropheus* spec. "Elongatus Aggressive"

Lake Malawi: Mumbo Isl., 10 cm
♂ Photo: A.Spreinat

A78210-5 *Pseudotropheus* spec. "Elongatus Aggressive"

Lake Malawi: Thumbi West Isl., 10 cm
♀ Photo: A.Spreinat

A78210-4 *Pseudotropheus* spec. "Elongatus Aggressive"

Lake Malawi, 10 cm
♂ Photo: W.Staeck

A78265-3 *Pseudotropheus flavus* Stauffer, 1988
before: *Pseudotropheus* spec. "Elongatus Dinghani"
Lake Malawi: Chinyakwazi, 12 cm
Photo: W.Staeck

A78395-4 *Pseudotropheus flavus* Stauffer, 1988

Lake Malawi: Chinyankwazi, 12 cm
♂ Photo: A.Spreinat

A78395-4 *Pseudotropheus flavus* Stauffer, 1988

Lake Malawi: Chinyankwazi, 12 cm
♀ Photo: A.Spreinat

A79030-4 *Pseudotropheus* spec. "Elongatus Yellow Tail"
=? *Pseudotropheus flavus* Stauffer, 1988
Lake Malawi: Mumbo Isl., 12 cm
♂ Photo: A.Spreinat

A79030-5 *Pseudotropheus* spec. "Elongatus Yellow Tail"
=? *Pseudotropheus flavus* Stauffer, 1988
Lake Malawi: Mumbo Isl., 12 cm
♂ Photo: W.Staeck

A78295-5 *Pseudotropheus* spec. "Elongatus Mbenji Blue"

Lake Malawi: Mbenji Isl., 10 cm

♂ Photo: E.Schraml

A78295-5 *Pseudotropheus* spec. "Elongatus Mbenji Blue"

Lake Malawi: Mbenji Isl., 10 cm

♀ Photo: E.Schraml

A78295-4 *Pseudotropheus* spec. "Elongatus Mbenji Blue"

Lake Malawi: Mbenji Isl., 10 cm

♂ Photo: A.Spreinat

A78295-5 *Pseudotropheus* spec. "Elongatus Mbenji Blue"

Lake Malawi: Mbenji Isl., 10 cm

♂ Photo: E.Schraml

A78225-5 *Pseudotropheus* spec. "Elongatus Brown"

Lake Malawi: Eccles Reef, 11 cm

♂ Photo: Ad Konings

A78225-4 *Pseudotropheus* spec. "Elongatus Brown"

Lake Malawi: Nakanthenga Isl., 11 cm

♂ Photo: A.Spreinat

A78275-4 *Pseudotropheus* spec. "Kirondo"

Lake Malawi: Kirondo (Tanzania), 10 cm

♂ Photo: Ad Konings

A78484-5 *Pseudotropheus* spec. "Elongatus Ruarwe"

Lake Malawi: Kakusa, 10 cm

♂ Photo: Ad Konings

A78220-5 *Pseudotropheus* spec. "Elongatus Boadzulu" leg. Spreinat
= *Pseudotropheus* spec. "Elongatus Greenback" leg. Konings
Lake Malawi: Boadzulu Isl., 9-10 cm
Photo: A.Spreinat

A78220-5 *Pseudotropheus* spec. "Elongatus Greenback" leg. Konings
= *Pseudotropheus* spec. "Elongatus Boadzulu" leg. Spreinat
Lake Malawi: Tsano Rock, 9-10 cm
Photo: Ad Konings

A78220-5 *Pseudotropheus* spec. "Elongatus Greenback" leg. Konings
= *Pseudotropheus* spec. "Elongatus Boadzulu" leg. Spreinat
Lake Malawi: Chemwezi Rock, 9-10 cm
Photo: Ad Konings

A78220-5 *Pseudotropheus* spec. "Elongatus Greenback" leg. Konings
= *Pseudotropheus* spec. "Elongatus Boadzulu" leg. Spreinat
Lake Malawi: Mumbo Isl., 9-10 cm
Photo: Ad Konings

A78365-5 *Pseudotropheus* spec. "Elongatus Masimbwe"

Lake Malawi: Likoma: Masimbwe Isl., 10-12 cm
Photo: A.Spreinat

A78365-5 *Pseudotropheus* spec. "Elongatus Masimbwe"

Lake Malawi: Likoma, 10-12 cm
Photo: Ad Konings

A78366-5 *Pseudotropheus* spec. "Elongatus Masimbwe"

Lake Malawi: Mara Point (Mozambique), 10-12 cm
Photo: Ad Konings

A78315-5 *Pseudotropheus* spec. "Elongatus Mpanga"

Lake Malawi: Chilumba, 10-12 cm
Photo: E.Schraml

A78315-5 *Pseudotropheus* spec. "Elongatus Mpanga"

Lake Malawi: Chilumba, 10-12 cm
♂ Photo: A.Spreinat

A78315-5 *Pseudotropheus* spec. "Elongatus Mpanga"

Lake Malawi: Chilumba, 10-12 cm
♀ Photo: A.Spreinat

A78315-4 *Pseudotropheus* spec. "Elongatus Mpanga"

Lake Malawi: Chilumba: Chirwa Isl., 10-12 cm
♂ Photo: A.Spreinat

A78270-5 *Pseudotropheus* spec. "Elongatus Gold Bar"

Lake Malawi: Likoma: Msekwa Point, 10-12 cm
♂ Photo: A.Spreinat

A78270-5 *Pseudotropheus* spec. "Elongatus Gold Bar"

Lake Malawi: Likoma: Khuyu, 10-12 cm
♂ Photo: A.Spreinat

A78270-5 *Pseudotropheus* spec. "Elongatus Gold Bar"

Lake Malawi: Chizumulu Isl., 10-12 cm
♂ Photo: Ad Konings

A78260-5 *Pseudotropheus* spec. "Elongatus Chitimba"

Lake Malawi: Chitimba Bay, ? cm
♂ Photo: Ad Konings

A78215-5 *Pseudotropheus* spec. "Elongatus Bee"

Lake Malawi: Chirwa Isl., 10-12 cm
♂ Photo: Ad Konings

A78216-5 *Pseudotropheus* spec. "Elongatus Bee"

Lake Malawi: Chilumba, Chitendi, 10-12 cm

♂ Photo: A.Spreinat

A78216-5 *Pseudotropheus* spec. "Elongatus Bee"

Lake Malawi, 10-12 cm

♂ Photo: W.Deproost

A78236-5 *Pseudotropheus* spec. "Elongatus Chailosi"

Lake Malawi: Chitande Isl., 10-12 cm

♂ Photo: Ad Konings

A78245-5 *Pseudotropheus* spec. "Chewere"

Lake Malawi: Chewere, 10-12 cm

♂ Photo: E.Schraml

A78245-4 *Pseudotropheus* spec. "Chewere"

Lake Malawi: Chewere, 10-12 cm

♀ Photo: A.Spreinat

A78245-5 *Pseudotropheus* spec. "Chewere"

Lake Malawi: Chewere, 10-12 cm

♀ Photo: E.Schraml

A78245-5 *Pseudotropheus* spec. "Chewere"

Lake Malawi: Chewere, 10-12 cm

♂ Photo: E.Schraml

A78250-4 *Pseudotropheus* spec. "Elongatus Chipoka"

Lake Malawi: Chidunga Rocks, ? cm

♂ Photo: Ad Konings

A78360-5 *Pseudotropheus* spec. "Elongatus Usisya"

Lake Malawi: Usisya, 10-12 cm

Photo: W.Deproost

A78360-5 *Pseudotropheus* spec. "Elongatus Usisya"

Lake Malawi: Usisya, 10-12 cm

Photo: W.Deproost

A78360-5 *Pseudotropheus* spec. "Elongatus Usisya"

Lake Malawi: Usisya, 10-12 cm

Photo: A.Spreinat

A78360-5 *Pseudotropheus* spec. "Elongatus Usisya"

Lake Malawi: Usisya, 10-12 cm

Photo: A.Spreinat

A78360-5 *Pseudotropheus* spec. "Elongatus Usisya"

Lake Malawi: Mara Rocks, 10-12 cm

Photo: Ad Konings

A78680-4 *Pseudotropheus* spec. "Staecki"
=? *Pseudotropheus* spec. "Elongatus Usisya"
Lake Malawi, 10-12 cm

Photo: E.Schraml

A78410-5 *Pseudotropheus (Tropheops) gracilior* TREWAVAS, 1935 leg. KONINGS, Typenfundort konträr
Lake Malawi: Tsano Rock, 12 cm

Photo: Ad Konings

A78410-5 *Pseudotropheus (Tropheops) gracilior* TREWAVAS, 1935 leg. KONINGS, Typenfundort konträr
Lake Malawi: Nkhudzi, 12 cm

Photo: Ad Konings

A78410-5 *Pseudotropheus (Tropheops) gracilior* Trewavas, 1935 leg. Konings, Typenfundort konträr
Lake Malawi: Otter Point, 12 cm
♂
Photo: Ad Konings

A78410-5 *Pseudotropheus (Tropheops) gracilior* Trewavas, 1935 leg. Konings, Typenfundort konträr
Lake Malawi: Otter Point, 12 cm
♀
Photo: Ad Konings

A78410-5 *Pseudotropheus (Tropheops) gracilior* Trewavas, 1935
Lake Malawi, 12 cm
♂
Photo: W.Staeck

A78600-5 *Pseudotropheus (Tropheops) novemfasciatus* Regan, 1922
Lake Malawi: Thumbi East Isl., 12 cm
♂
Photo: Ad Konings

A78711-4 *Pseudotropheus (Tropheops) tropheops* Regan, 1922
Lake Malawi: Makokola Reef, 13 cm
♂
Photo: Ad Konings

A78711-4 *Pseudotropheus (Tropheops) tropheops* Regan, 1922
Lake Malawi: Makokola Reef, 13 cm
♀
Photo: Ad Konings

A78711-5 *Pseudotropheus (Tropheops) tropheops* Regan, 1922
OB-Morph
Lake Malawi: Makokola Reef, 13 cm
♂
Photo: Ad Konings

A78711-5 *Pseudotropheus (Tropheops) tropheops* Regan, 1922
OB-Morph
Lake Malawi: Makokola Reef, 13 cm
♀
Photo: Ad Konings

A78717-5 *Pseudotropheus (Tropheops) tropheops* Regan, 1922

Lake Malawi: Zimbawe Rock, 13 cm

Photo: Ad Konings

A78717-5 *Pseudotropheus (Tropheops) tropheops* Regan, 1922

Lake Malawi: Zimbawe Rock, 13 cm

Photo: Ad Konings

A78718-4 *Pseudotropheus (Tropheops) tropheops* Regan, 1922

Lake Malawi: Tsano Rock, 13 cm

Photo: Ad Konings

A78716-5 *Pseudotropheus (Tropheops) tropheops* Regan, 1922

Lake Malawi: Monkey Bay, 13 cm

Photo: W.Staeck

A78843-5 *Pseudotropheus (Tropheops)* spec. "Tropheops Aurora" leg. Konings, = *Ps. (T.)* spec. "Tropheops Mbamba" leg. Spreinat
Lake Malawi: Njambe (Tanzania), 13 cm

Photo: Ad Konings

A78845-5 *Pseudotropheus (Tropheops)* spec. "Tropheops Aurora" leg. Konings, = *Ps. (T.)* spec. "Tropheops Mbamba" leg. Spreinat
Lake Malawi: Tumbi Point (Tanzania), 13 cm

Photo: Ad Konings

A78846-5 *Pseudotropheus (Tropheops)* spec. "Tropheops Mbamba" leg. Spreinat, = *Ps. (T.)* spec. "Tropheops Aurora" leg. Konings
Lake Malawi: Higga Reef (Tanzania), 13 cm

Photo: A.Spreinat

A78846-5 *Pseudotropheus (Tropheops)* spec. "Tropheops Mbamba" leg. Spreinat, = *Ps. (T.)* spec. "Tropheops Aurora" leg. Konings
Lake Malawi: Higga Reef (Tanzania), 13 cm

Photo: A.Spreinat

A78844-5 *Pseudotropheus (Tropheops)* spec. "Tropheops Mbamba" leg. SPREINAT, = *Ps. (T.)* spec. "Tropheops Aurora" leg. KONINGS
Lake Malawi: Puulu Isl. (Tanzania), 13 cm
♂ Photo: A.Spreinat

A78844-5 *Pseudotropheus (Tropheops)* spec. "Tropheops Mbamba" leg. SPREINAT, = *Ps. (T.)* spec. "Tropheops Aurora" leg. KONINGS
Lake Malawi: Puulu Isl. (Tanzania), 13 cm
♀ Photo: A.Spreinat

A78847-5 *Pseudotropheus (Tropheops)* spec. "Tropheops Mbamba" leg. SPREINAT, = *Ps. (T.)* spec. "Tropheops Aurora" leg. KONINGS
Lake Malawi: Lundo Isl. (Tanzania), 13 cm
♂ Photo: A.Spreinat

A78841-4 *Pseudotropheus (Tropheops)* spec. "Tropheops Mbamba" leg. SPREINAT, = *Ps. (T.)* spec. "Tropheops Aurora" leg. KONINGS
Lake Malawi: Mbamba Bay (Tanzania): Mara Rocks, 13 cm
Photo: A.Spreinat

A78860-5 *Pseudotropheus (Tropheops)* spec. "Tropheops Membe"

Lake Malawi: Likoma Isl.: Membe Point, 13 cm
♂ Photo: A.Spreinat

A78860-5 *Pseudotropheus (Tropheops)* spec. "Tropheops Membe"

Lake Malawi: Likoma Isl.: Mbuzi Isl., 13 cm
♀ Photo: A.Spreinat

A78860-5 *Pseudotropheus (Tropheops)* spec. "Tropheops Membe"

Lake Malawi: Likoma Isl.: Makulawe Point, 13 cm
♂ Photo: A.Spreinat

A78860-5 *Pseudotropheus (Tropheops)* spec. "Tropheops Membe"

Lake Malawi: Likoma Isl.: Makulawe Point, 13 cm
♀ Photo: A.Spreinat

A78861-5 *Pseudotropheus (Tropheops)* spec. "Tropheops Membe"

Lake Malawi: Cobue (Mozambique): Mara Point, 13 cm

Photo: A.Spreinat

A78861-5 *Pseudotropheus (Tropheops)* spec. "Tropheops Membe"

Lake Malawi: Cobue (Mozambique): Mara Point, 13 cm

Photo: A.Spreinat

A78861-5 *Pseudotropheus (Tropheops)* spec. "Tropheops Membe"

Lake Malawi: Cobue (Mozambique): Mara Point, 13 cm

Photo: Ad Konings

A78861-5 *Pseudotropheus (Tropheops)* spec. "Tropheops Membe"

Lake Malawi: Tumbi Point (Mozambique), 13 cm

Photo: Ad Konings

A78861-5 *Pseudotropheus (Tropheops)* spec. "Tropheops Membe"

Lake Malawi: Mbweca (Mozambique): Mara Point, 13 cm

Photo: Ad Konings

A78861-5 *Pseudotropheus (Tropheops)* spec. "Tropheops Membe"

Lake Malawi: Mbweca (Mozambique): Mara Point, 13 cm

Photo: Ad Konings

A78868-5 *Pseudotropheus (Tropheops)* spec. "Tropheops Mauve Yellow"

Lake Malawi: Magunga (Tanzania), 13 cm

Photo: Ad Konings

A78868-5 *Pseudotropheus (Tropheops)* spec. "Tropheops Mauve Yellow"

Lake Malawi: Magunga (Tanzania), 13 cm

Photo: Ad Konings

A78867-5 *Pseudotropheus (Tropheops)* spec. "Tropheops Mauve Yellow"

Lake Malawi: Cape Kaiser (Tanzania), 13 cm
♂ Photo: Ad Konings

A78838-5 *Pseudotropheus (Tropheops)* spec. "Tropheops Mauve"

Lake Malawi: Mara Rocks, 10 cm
♂ Photo: Ad Konings

A78836-5 *Pseudotropheus (Tropheops)* spec. "Tropheops Mauve"

Lake Malawi: Chewere, 10 cm
♂ Photo: Ad Konings

A78836-5 *Pseudotropheus (Tropheops)* spec. "Tropheops Mauve"

Lake Malawi: Chewere, 10 cm
♀ Photo: Ad Konings

A78839-5 *Pseudotropheus (Tropheops)* spec. "Tropheops Mauve"

Lake Malawi: Usisya, 10 cm
♂ Photo: A.Spreinat

A78837-5 *Pseudotropheus (Tropheops)* spec. "Tropheops Mauve"

Lake Malawi: Chilumba: Chitendi Isl., 10 cm
♂ Photo: A.Spreinat

A78835-4 *Pseudotropheus (Tropheops)* spec. "Tropheops Mauve"

Aquarientiere/Aquarium specimen
♂ Photo: A.Spreinat

A78835-4 *Pseudotropheus (Tropheops)* spec. "Tropheops Mauve"

Aquarientiere/Aquarium specimen
♀ Photo: A.Spreinat

A78825-5 *Pseudotropheus (Tropheops)* spec. "Tropheops Maleri Yellow"

Lake Malawi: Namalenje Isl., 12 cm
♂
Photo: W.Deproost

A78825-5 *Pseudotropheus (Tropheops)* spec. "Tropheops Maleri Yellow"

Lake Malawi: Namalenje Isl., 12 cm
♀
Photo: Ad Konings

A78825-5 *Pseudotropheus (Tropheops)* spec. "Tropheops Maleri Yellow"

Lake Malawi: Maleri Isl., 12 cm
♂
Photo: Ad Konings

A78825-5 *Pseudotropheus (Tropheops)* spec. "Tropheops Maleri Yellow"

Lake Malawi: Nakanthenga Isl., 12 cm
♂
Photo: A.Spreinat

A78855-5 *Pseudotropheus (Tropheops)* spec. "Tropheops Mbenji Yellow"

Lake Malawi: Mbenji Isl., 12 cm
♂
Photo: Ad Konings

A78855-5 *Pseudotropheus (Tropheops)* spec. "Tropheops Mbenji Yellow"

Lake Malawi: Mbenji Isl., 12 cm
♂ ♀
Photo: Ad Konings

A78807-5 *Pseudotropheus (Tropheops)* spec. "Tropheops Lilac Mumbo"

Lake Malawi: Mumbo Isl., 13 cm
♂
Photo: A.Spreinat

A78807-4 *Pseudotropheus (Tropheops)* spec. "Tropheops Lilac Mumbo"

Lake Malawi: Mumbo Isl., 13 cm
♀
Photo: A.Spreinat

A78806-5 *Pseudotropheus (Tropheops)* spec. "Tropheops Lilac"

Lake Malawi: Nakantenga Isl., 12 cm

♂ Photo: Ad Konings

A78808-5 *Pseudotropheus (Tropheops)* spec. "Tropheops Lilac"

Lake Malawi: Chidunga Rocks, 12 cm

♂ Photo: Ad Konings

A78809-5 *Pseudotropheus (Tropheops)* spec. "Tropheops Lilac"

Lake Malawi: Thumbi West Isl., 12 cm

♂ Photo: Ad Konings

A78791-5 *Pseudotropheus (Tropheops)* spec. "Tropheops Gold Breast"

Lake Malawi: Tumbi Point (Mozambique), 13 cm

♂ Photo: Ad Konings

A78792-5 *Pseudotropheus (Tropheops)* spec. "Tropheops Gold Breast"

Lake Malawi: Londo (Mozambique), 13 cm

♂ Photo: Ad Konings

A78792-5 *Pseudotropheus (Tropheops)* spec. "Tropheops Gold Breast"

Lake Malawi: Londo (Mozambique), 13 cm

♀ Photo: Ad Konings

A78793-5 *Pseudotropheus (Tropheops)* spec. "Tropheops Gold Breast"

Lake Malawi: Minga Bay (Mozambique): Ngumbe Rock, 13 cm

♂ Photo: A.Spreinat

A78794-4 *Pseudotropheus (Tropheops)* spec. "Tropheops Gold Breast"

Lake Malawi: Ponta Messuli (Mozambique), 13 cm

♂ Photo: A.Spreinat

A78785-5 *Pseudotropheus (Tropheops)* spec. "Tropheops Deep"

Lake Malawi: Nkhata Bay, 11 cm

♂ Photo: Ad Konings

A78785-5 *Pseudotropheus (Tropheops)* spec. "Tropheops Deep"

Lake Malawi: Nkhata Bay, 11 cm

♀ Photo: Ad Konings

A78786-5 *Pseudotropheus (Tropheops)* spec. "Tropheops Deep"

Lake Malawi: Lion's Cove, 11 cm

♂ Photo: Ad Konings

A78786-3 *Pseudotropheus (Tropheops)* spec. "Tropheops Deep"

Lake Malawi: Lion's Cove, 11 cm

♀ Photo: Ad Konings

A78780-5 *Pseudotropheus (Tropheops)* spec. "Tropheops Dark" leg. KONINGS = *Ps. (T.)* spec. "Tropheops Membe" leg. SPREINAT

Lake Malawi: Likoma Isl., 14 cm

♂ Photo: Ad Konings

A78780-5 *Pseudotropheus (Tropheops)* spec. "Tropheops Dark"

Lake Malawi: Likoma Isl., 14 cm

♀ Photo: Ad Konings

A78780-5 *Pseudotropheus (Tropheops)* spec. "Tropheops Dark"

Lake Malawi: Likoma Isl.: Ndumbi Rocks, 14 cm

♂ Photo: A.Spreinat

A78780-5 *Pseudotropheus (Tropheops)* spec. "Tropheops Dark"

Lake Malawi: Likoma Isl.: Khuyu, 14 cm

♂ Photo: A.Spreinat

A78730-5 *Pseudotropheus (Tropheops)* spec. "Tropheops Black"

Lake Malawi: Chilumba: Chitendi Isl., 13 cm

Photo: A.Spreinat

A78730-5 *Pseudotropheus (Tropheops)* spec. "Tropheops Black"

Lake Malawi: Chilumba: Chitendi Isl., 13 cm

Photo: A.Spreinat

A78730-5 *Pseudotropheus (Tropheops)* spec. "Tropheops Black"

Lake Malawi: Chilumba: Mpanga Rocks, 13 cm

Photo: A.Spreinat

A78726-5 *Pseudotropheus (Tropheops)* spec. "Tropheops Black"

Lake Malawi: Chesese, 13 cm

Photo: Ad Konings

A78729-5 *Pseudotropheus (Tropheops)* spec. "Tropheops Black"

Lake Malawi: Mbowe Isl., 13 cm

Photo: Ad Konings

A78733-5 *Pseudotropheus (Tropheops)* spec. "Tropheops Black"

Lake Malawi: Magunga (Tanzania), 13 cm

Photo: Ad Konings

A78731-5 *Pseudotropheus (Tropheops)* spec. "Tropheops Black"

Lake Malawi: Makonde (Tanzania), 13 cm

Photo: Ad Konings

A78731-5 *Pseudotropheus (Tropheops)* spec. "Tropheops Black"
OB-Morph

Lake Malawi: Makonde (Tanzania), 13 cm

Photo: Ad Konings

A78727-5 *Pseudotropheus (Tropheops)* spec. "Tropheops Big Blue Yellow" leg. Spreinat =? *Ps. (T.)* spec. "Tropheops Black" leg. Konings
Lake Malawi: Lundo Isl. (Tanzania), 13 cm
Photo: A.Spreinat

A78734-5 *Pseudotropheus (Tropheops)* spec. "Tropheops Big Blue Yellow" leg. Spreinat =? *Ps. (T.)* spec. "Tropheops Black" leg. Konings
Lake Malawi: Tumbi Rocks (Tanzania), 13 cm
Photo: A.Spreinat

A78728-5 *Pseudotropheus (Tropheops)* spec. "Tropheops Black" leg. Konings=? *Ps. (T.)* spec. "Tropheops Big Blue Yellow" leg. Spreinat
Lake Malawi: Lundu (Tanzania), 13 cm
Photo: Ad Konings

A78732-5 *Pseudotropheus (Tropheops)* spec. "Tropheops Black" leg. Konings=? *Ps. (T.)* spec. "Tropheops Big Blue Yellow" leg. Spreinat
Lake Malawi: Lupingu (Tanzania), 13 cm
Photo: Ad Konings

A78720-5 *Pseudotropheus (Tropheops)* spec. "Tropheops Band"

Lake Malawi: Nkhata Bay, 13 cm
Photo: Ad Konings

A78720-5 *Pseudotropheus (Tropheops)* spec. "Tropheops Band"

Lake Malawi: Nkhata Bay, 13 cm
Photo: Ad Konings

A78850-5 *Pseudotropheus (Tropheops)* spec. "Tropheops Mbenji Blue"

Lake Malawi: Mbenji Isl., 12 cm
Photo: Ad Konings

A78850-5 *Pseudotropheus (Tropheops)* spec. "Tropheops Mbenji Blue"

Lake Malawi: Mbenji Isl., 12 cm
Photo: Ad Konings

A78851-5 *Pseudotropheus (Tropheops)* spec. "Tropheops Mbenji Blue"

Lake Malawi: Jalo Reef, 12 cm

Photo: Ad Konings

A78852-5 *Pseudotropheus (Tropheops)* spec. "Tropheops Mbenji Blue"

Lake Malawi: Nkhomo Reef, 12 cm

Photo: Ad Konings

A78870-5 *Pseudotropheus (Tropheops)* spec. "Tropheops No Band"

Lake Malawi: Nkhata Bay, 12 cm

Photo: Ad Konings

A78870-5 *Pseudotropheus (Tropheops)* spec. "Tropheops No Band"

Lake Malawi: Nkhata Bay, 12 cm

Photo: Ad Konings

A78942-5 *Pseudotropheus (Tropheops)* spec. "Tropheops Yellow Chin"

Lake Malawi: Likoma: Ndomo Point, 14 cm

Photo: A.Spreinat

A78942-5 *Pseudotropheus (Tropheops)* spec. "Tropheops Yellow Chin"

Lake Malawi: Likoma: Mbuzi Isl., 14 cm

Photo: W.Deproost

A78942-5 *Pseudotropheus (Tropheops)* spec. "Tropheops Yellow Chin"

Lake Malawi: Likoma, 14 cm

Photo: Ad Konings

A78942-5 *Pseudotropheus (Tropheops)* spec. "Tropheops Yellow Chin"

Lake Malawi: Likoma: Khuyu, 14 cm

Photo: A.Spreinat

A78941-5 *Pseudotropheus (Tropheops)* spec. "Tropheops Yellow Chin"

Lake Malawi: Mara Point (Mozambique), 14 cm

Photo: Ad Konings

A78941-5 *Pseudotropheus (Tropheops)* spec. "Tropheops Yellow Chin"

Lake Malawi: Mara Point (Mozambique), 14 cm

Photo: Ad Konings

A78941-5 *Pseudotropheus (Tropheops)* spec. "Tropheops Yellow Chin"

Lake Malawi: Mara Point (Mozambique), 14 cm

Photo: Ad Konings

A78943-5 *Pseudotropheus (Tropheops)* spec. "Tropheops Yellow Chin"

Lake Malawi: Liutche (Mozambique), 14 cm

Photo: Ad Konings

A78940-5 *Pseudotropheus (Tropheops)* spec. "Tropheops Yellow Chin"

Lake Malawi: Chisumulu: Membe Isl., 14 cm

Photo: A.Spreinat

A78940-5 *Pseudotropheus (Tropheops)* spec. "Tropheops Yellow Chin"

Lake Malawi: Chisumulu: Chiteko, 14 cm

Photo: A.Spreinat

A78951-5 *Pseudotropheus (Tropheops)* spec. "Tropheops Yellow Chin" leg. Konings = *Ps. (T.)* spec. "Tropheops Yellow Head" leg. Spreinat

Lake Malawi: Hai Reef (Tanzania), 14 cm

Photo: Ad Konings

A78951-5 *Pseudotropheus (Tropheops)* spec. "Tropheops Yellow Chin" leg. Konings = *Ps. (T.)* spec. "Tropheops Yellow Head" leg. Spreinat

Lake Malawi: Hai Reef (Tanzania), 14 cm

Photo: A.Spreinat

A78954-5 *Pseudotropheus (Tropheops)* spec. "Tropheops Yellow Head"

Lake Malawi: Undu Point (Tanzania), 14 cm

Photo: A.Spreinat

A78953-5 *Pseudotropheus (Tropheops)* spec. "Tropheops Yellow Head"

Lake Malawi: Tumbi Rocks (Tanzania), 14 cm

Photo: A.Spreinat

A78945-5 *Pseudotropheus (Tropheops)* spec. "Tropheops Yellow Gular"

Lake Malawi: Eccles Reef, 13 cm

Photo: Ad Konings

A78945-5 *Pseudotropheus (Tropheops)* spec. "Tropheops Yellow Gular"

Lake Malawi: Eccles Reef, 13 cm

Photo: Ad Konings

A78946-5 *Pseudotropheus (Tropheops)* spec. "Tropheops Yellow Gular"

Lake Malawi: Ntekete, 13 cm

Photo: Ad Konings

A78947-5 *Pseudotropheus (Tropheops)* spec. "Tropheops Yellow Gular"

Lake Malawi: Gome, 13 cm

Photo: Ad Konings

A78948-5 *Pseudotropheus (Tropheops)* spec. "Tropheops Yellow Gular"

Lake Malawi: Fort Maguire, 13 cm

Photo: A.Spreinat

A78949-5 *Pseudotropheus (Tropheops)* spec. "Tropheops Yellow Gular"

Lake Malawi: Tumbi Point (Mozambique), 13 cm

Photo: Ad Konings

A78883-5 *Pseudotropheus (Tropheops)* spec. "Tropheops Red Cheek"
Lake Malawi: Likoma: Maingano, 12-13 cm
Photo: W.Deproost

A78883-5 *Pseudotropheus (Tropheops)* spec. "Tropheops Red Cheek"
Lake Malawi: Monkey Bay: Tsano Rock, 12-13 cm
Photo: A.Spreinat

A78885-5 *Pseudotropheus (Tropheops)* spec. "Tropheops Red Cheek"
Lake Malawi: Thumbi West Isl., 12-13 cm
Photo: A.Spreinat

A78886-5 *Pseudotropheus (Tropheops)* spec. "Tropheops Red Cheek"
Lake Malawi: Chisumulu: Membe Isl., 12-13 cm
Photo: A.Spreinat

A78886-5 *Pseudotropheus (Tropheops)* spec. "Tropheops Red Cheek"
Lake Malawi: Chisumulu: Chiwe Rocks, 12-13 cm
Photo: A.Spreinat

A78886-5 *Pseudotropheus (Tropheops)* spec. "Tropheops Red Cheek"
Lake Malawi: Chisumulu: Chiwe Rocks, 12-13 cm
Photo: A.Spreinat

A78887-5 *Pseudotropheus (Tropheops)* spec. "Tropheops Red Cheek" leg. KONINGS = *Ps. (T.)* spec. "Tropheops Checkered" leg. SPREINAT
Lake Malawi: Makonde (Tanzania), 12-13 cm
Photo: Ad Konings

A78888-5 *Pseudotropheus (Tropheops)* spec. "Tropheops Red Cheek" leg. KONINGS = *Ps. (T.)* spec. "Tropheops Checkered" leg. SPREINAT
Lake Malawi: Lupingu (Tanzania), 12-13 cm
Photo: Ad Konings

A78881-5 *Pseudotropheus (Tropheops)* spec. "Tropheops Red Cheek" leg. KONINGS = *Ps. (T.)* spec. "Tropheops Checkered" leg. SPREINAT
Lake Malawi: Ikombe (Tanzania), 12-13 cm
Photo: Ad Konings

A78881-5 *Pseudotropheus (Tropheops)* spec. "Tropheops Red Cheek" leg. KONINGS = *Ps. (T.)* spec. "Tropheops Checkered" leg. SPREINAT
Lake Malawi: Ikombe (Tanzania), 12-13 cm
Photo: Ad Konings

A78884-5 *Pseudotropheus (Tropheops)* spec. "Tropheops Red Cheek" leg. KONINGS = *Ps. (T.)* spec. "Tropheops Checkered" leg. SPREINAT
Lake Malawi: Magunga (Tanzania), 12-13 cm
Photo: Ad Konings

A78884-5 *Pseudotropheus (Tropheops)* spec. "Tropheops Red Cheek" leg. KONINGS = *Ps. (T.)* spec. "Tropheops Checkered" leg. SPREINAT
Lake Malawi: Magunga (Tanzania), 12-13 cm
Photo: Ad Konings

A78747-5 *Pseudotropheus (Tropheops)* spec. "Tropheops Checkered" leg. SPREINAT =? *Ps. (T.)* spec. "Tropheops Red Cheek" leg. KONINGS
Lake Malawi: Lumbira (Tanzania), 10-12 cm
Photo: A.Spreinat

A78747-5 *Pseudotropheus (Tropheops)* spec. "Tropheops Checkered" leg. SPREINAT =? *Ps. (T.)* spec. "Tropheops Red Cheek" leg. KONINGS
Lake Malawi: Nkanda (Tanzania), 10-12 cm
Photo: A.Spreinat

A78745-5 *Pseudotropheus (Tropheops)* spec. "Tropheops Broad Mouth"

Lake Malawi: Thumbi West Isl., 12 cm
Photo: A.Spreinat

A78745-5 *Pseudotropheus (Tropheops)* spec. "Tropheops Broad Mouth" ? OB-Morph
Lake Malawi, 12 cm
Photo: W.Staeck

Felsküste im Süden des Malawisees
Rocky coast in the south of Lake Malawi

Photo: W.Staeck

A78865-5 *Pseudotropheus (Tropheops)* spec. "Tropheops Mutant" =? *Ps. (T.)* spec. "Tropheops Chitande Yellow"
Lake Malawi: Cove Mountain (Tanzania), 11-13 cm

Photo: A.Spreinat

A78865-5 *Pseudotropheus (Tropheops)* spec. "Tropheops Mutant" =? *Ps. (T.)* spec. "Tropheops Chitande Yellow"
Lake Malawi: Cove Mountain (Tanzania), 11-13 cm

Photo: A.Spreinat

A78772-5 *Pseudotropheus (Tropheops)* spec. "Tropheops Chitande Yellow" leg. Spreinat =? *Ps. (T.) macrophthalmus* Ahl, 1927 leg. Konings
Lake Malawi: Kirondo (Tanzania), 11-13 cm

Photo: A.Spreinat

A78519-5 *Pseudotropheus (Tropheops) macrophthalmus* Ahl, 1927 leg. Konings =? *Ps. (T.)* spec. "Tropheops Chitande Yellow" leg. Spreinat
Lake Malawi: Kirondo (Tanzania), 11-13 cm

Photo: Ad Konings

A78773-5 *Pseudotropheus (Tropheops)* spec. "Tropheops Chitande Yellow" leg. Spreinat =? *Ps. (T.) macrophthalmus* Ahl, 1927 leg. Konings
Lake Malawi: Chilumba: Chirwa Isl., 11-13 cm
Photo: A.Spreinat

A78771-5 *Pseudotropheus (Tropheops)* spec. "Tropheops Chitande Yellow" leg. Spreinat =? *Ps. (T.) macrophthalmus* Ahl, 1927 leg. Konings
Lake Malawi: Ikombe (Tanzania), 11-13 cm
Photo: A.Spreinat

A78535-5 *Pseudotropheus (Tropheops)* spec. "Macrophthalmus Chitimba" =? *Ps. (T.) macrophthalmus* Ahl, 1927
Lake Malawi: Chitimba Bay, 11-13 cm
Photo: Ad Konings

A78535-5 *Pseudotropheus (Tropheops)* spec. "Macrophthalmus Chitimba" =? *Ps. (T.) macrophthalmus* Ahl, 1927
Lake Malawi: Chitimba Bay, 11-13 cm
Photo: Ad Konings

A78517-5 *Pseudotropheus (Tropheops) macrophthalmus* Ahl, 1927 leg. Konings = *Tropheops* spec. "Tropheops Chitande Yellow" leg. Spreinat
Lake Malawi: Chitande Isl., 11-13 cm
Photo: Ad Konings

A78518-5 *Pseudotropheus (Tropheops) macrophthalmus* Ahl, 1927 leg. Konings = *Tropheops* spec. "Tropheops Chitande Yellow" leg. Spreinat
Lake Malawi: Chizumulu Isl., 11-13 cm
Photo: Ad Konings

A78516-5 *Pseudotropheus (Tropheops) macrophthalmus* Ahl, 1927 leg. Konings
Lake Malawi: Chiloelo (Mozambique), 11-13 cm
Photo: Ad Konings

A78516-5 *Pseudotropheus (Tropheops) macrophthalmus* Ahl, 1927 leg. Konings
Lake Malawi: Chiloelo (Mozambique), 11-13 cm
Photo: Ad Konings

A78797-5 *Pseudotropheus (Tropheops)* spec. "Tropheops Gold"
=? *Ps. (T.) macrophthalmus* Ahl, 1927
Lake Malawi: Chisumulu: Membe Isl., 11-13 cm
Photo: A.Spreinat

A78798-5 *Pseudotropheus (Tropheops)* spec. "Tropheops Gold"
=? *Ps. (T.) macrophthalmus* Ahl, 1927
Lake Malawi: Chikala (Mozambique), 11-13 cm
Photo: A.Spreinat

A78799-5 *Pseudotropheus (Tropheops)* spec. "Tropheops Gold"
=? *Ps. (T.) macrophthalmus* Ahl, 1927
Lake Malawi: Cobue (Mozambique): Kanjindo, 11-13 cm
Photo: A.Spreinat

A78799-5 *Pseudotropheus (Tropheops)* spec. "Tropheops Gold"
=? *Ps. (T.) macrophthalmus* Ahl, 1927
Lake Malawi: Cobue (Mozambique): Kanjindo, 11-13 cm
Photo: A.Spreinat

A78796-5 *Pseudotropheus (Tropheops)* spec. "Tropheops Gold"
=? *Ps. (T.) macrophthalmus* Ahl, 1927
Lake Malawi: Metangula (Mozambique): Chuanga, 11-13 cm
Photo: A.Spreinat

A78795-5 *Pseudotropheus (Tropheops)* spec. "Tropheops Gold Otter"
Lake Malawi: Otter Point, 11-13 cm
Photo: Ad Konings

A78795-5 *Pseudotropheus (Tropheops)* spec. "Tropheops Gold Otter"
Lake Malawi: Otter Point, 11-13 cm
Photo: Ad Konings

A78878-5 *Pseudotropheus (Tropheops)* spec. "Tropheops Olive"
Lake Malawi: Usisya, 10-12 cm
Photo: A.Spreinat

A78875-5 *Pseudotropheus (Tropheops)* spec. "Tropheops Olive"

Lake Malawi, 10-12 cm
♂ Photo: W.Deproost

A78875-5 *Pseudotropheus (Tropheops)* spec. "Tropheops Olive"

Lake Malawi, 10-12 cm
♂ Photo: W.Deproost

A78875-5 *Pseudotropheus (Tropheops)* spec. "Tropheops Olive"

Lake Malawi, 10-12 cm
♀ Photo: W.Deproost

A78879-5 *Pseudotropheus (Tropheops)* spec. "Tropheops Olive"

Lake Malawi: Kirondo (Tanzania), 10-12 cm
♂ Photo: A.Spreinat

A78879-5 *Pseudotropheus (Tropheops)* spec. "Tropheops Olive"

Lake Malawi: Chitande Isl., 10-12 cm
♂ Photo: Ad Konings

A78879-5 *Pseudotropheus (Tropheops)* spec. "Tropheops Olive"

Lake Malawi: Chitande Isl., 10-12 cm
♀ Photo: Ad Konings

A78876-5 *Pseudotropheus (Tropheops)* spec. "Tropheops Olive"

Lake Malawi: Cape Kaiser (Tanzania), 10-12 cm
♂ Photo: Ad Konings

A78874-5 *Pseudotropheus (Tropheops)* spec. "Tropheops Olive"

Lake Malawi: Magunga (Tanzania), 10-12 cm
♀ Photo: Ad Konings

A78873-5 *Pseudotropheus (Tropheops)* spec. "Tropheops Olive"

Lake Malawi: Mdoka, 10-12 cm

♂ Photo: Ad Konings

A78872-5 *Pseudotropheus (Tropheops)* spec. "Tropheops Olive"

Lake Malawi: Lion's Cove, 10-12 cm

♂ Photo: Ad Konings

A78891-5 *Pseudotropheus (Tropheops)* spec. "Tropheops Red Fin"

Lake Malawi: Hai Reef (Tanzania), 10-13 cm

♂ Photo: A.Spreinat

A78891-3 *Pseudotropheus (Tropheops)* spec. "Tropheops Red Fin"

Lake Malawi: Hai Reef (Tanzania), 10-13 cm

♂ Photo: Ad Konings

A78892-5 *Pseudotropheus (Tropheops)* spec. "Tropheops Red Fin"

Lake Malawi: Hongi Isl. (Tanzania), 10-13 cm

♂ Photo: A.Spreinat

A78892-5 *Pseudotropheus (Tropheops)* spec. "Tropheops Red Fin"

Lake Malawi: Hongi Isl. (Tanzania), 10-13 cm

♀ Photo: A.Spreinat

A78892-5 *Pseudotropheus (Tropheops)* spec. "Tropheops Red Fin"

Lake Malawi: Hongi Isl. (Tanzania), 10-13 cm

♂ Photo: Ad Konings

A78867-5 *Pseudotropheus (Tropheops)* spec. "Tropheops Red Fin"

Lake Malawi: Undu Point (Tanzania), 10-13 cm

♂ Photo: A.Spreinat

A78894-5 *Pseudotropheus (Tropheops)* spec. "Tropheops Red Fin"

Lake Malawi: Lundu (Tanzania), 10-13 cm
♂ Photo: Ad Konings

A78894-2 *Pseudotropheus (Tropheops)* spec. "Tropheops Red Fin"

Lake Malawi: Lundu (Tanzania), 10-13 cm
Photo: Ad Konings

A78898-5 *Pseudotropheus (Tropheops)* spec. "Tropheops Red Fin"

Lake Malawi: Tumbi Point (Tanzania), 10-13 cm
♂ Photo: Ad Konings

A78898-4 *Pseudotropheus (Tropheops)* spec. "Tropheops Red Fin"

Lake Malawi: Tumbi Point (Tanzania), 10-13 cm
♀ Photo: Ad Konings

A78899-5 *Pseudotropheus (Tropheops)* spec. "Tropheops Red Fin"

Lake Malawi: Puulu (Tanzania), 10-13 cm
♂ Photo: Ad Konings

A78866-5 *Pseudotropheus (Tropheops)* spec. "Tropheops Red Fin"

Lake Malawi: Liuli (Tanzania), 10-13 cm
♂ Photo: Ad Konings

A78895-5 *Pseudotropheus (Tropheops)* spec. "Tropheops Red Fin"

Lake Malawi: Mbamba Bay (Tanzania): Luhuchi Rocks, 10-13 cm
♂ Photo: A.Spreinat

A78895-5 *Pseudotropheus (Tropheops)* spec. "Tropheops Red Fin"

Lake Malawi: Mbamba Bay (Tanzania): Luhuchi Rocks, 10-13 cm
♀ Photo: A.Spreinat

A78895-5 *Pseudotropheus (Tropheops)* spec. "Tropheops Red Fin"

Lake Malawi: Mbamba Bay (Tanzania): Mara Rocks, 10-13 cm
Photo: A.Spreinat

A78895-5 *Pseudotropheus (Tropheops)* spec. "Tropheops Red Fin"

Lake Malawi: Mbamba Bay (Tanzania): Mara Rocks, 10-13 cm
Photo: A.Spreinat

A78896-5 *Pseudotropheus (Tropheops)* spec. "Tropheops Red Fin"

Lake Malawi: Njambe (Tanzania), 10-13 cm
Photo: A.Spreinat

A78896-5 *Pseudotropheus (Tropheops)* spec. "Tropheops Red Fin"

Lake Malawi: Njambe (Tanzania), 10-13 cm
Photo: A.Spreinat

A78868-5 *Pseudotropheus (Tropheops)* spec. "Tropheops Red Fin"

Lake Malawi: Usisya (Malawi): Mara Rocks, 10-13 cm
Photo: A.Spreinat

A78869-5 *Pseudotropheus (Tropheops)* spec. "Tropheops Red Fin"

Lake Malawi: Kakusa, 10-13 cm
Photo: Ad Konings

A78814-5 *Pseudotropheus (Tropheops)* spec. "Tropheops Red Fin"

Lake Malawi: Hora Mhango, 10-13 cm
Photo: Ad Konings

A78890-5 *Pseudotropheus (Tropheops)* spec. "Tropheops Red Fin"

Lake Malawi: Wikihi (Mozambique), 10-13 cm
Photo: Ad Konings

A78893-5 *Pseudotropheus (Tropheops)* spec. "Tropheops Red Fin"

Lake Malawi: Londo (Mozambique), 10-13 cm

♂ Photo: A.Spreinat

A78800-5 *Pseudotropheus (Tropheops)* spec. "Tropheops Intermediate"
Aquariumspecimen

Lake Malawi: Thumbi West Isl., 11 cm

♂ Photo: A.Spreinat

A78800-5 *Pseudotropheus (Tropheops)* spec. "Tropheops Intermediate"

Lake Malawi: Thumbi West Isl., 11 cm

♂ Photo: Ad Konings

A78800-5 *Pseudotropheus (Tropheops)* spec. "Tropheops Intermediate"

Lake Malawi: Thumbi West Isl., 11 cm

♀ Photo: A.Spreinat

A78801-5 *Pseudotropheus (Tropheops)* spec. "Tropheops Intermediate"

Lake Malawi: Tsano Rock., 11 cm

♂ Photo: A.Spreinat

A78801-5 *Pseudotropheus (Tropheops)* spec. "Tropheops Intermediate"

Lake Malawi: Tsano Rock., 11 cm

♀ Photo: A.Spreinat

A78735-5 *Pseudotropheus (Tropheops)* spec. "Tropheops Black Dorsal"

Lake Malawi: Eccles Reef, 12-13 cm

♂ Photo: A.Spreinat

A78735-5 *Pseudotropheus (Tropheops)* spec. "Tropheops Black Dorsal"

Lake Malawi: Eccles Reef, 12-13 cm

♀ Photo: A.Spreinat

A78736-5 *Pseudotropheus (Tropheops)* spec. "Tropheops Black Dorsal"

Lake Malawi: Makanjila, 12-13 cm

♂ Photo: Ad Konings

A78765-5 *Pseudotropheus (Tropheops)* spec. "Tropheops Chinyankwazi"

Lake Malawi: Chinyankwazi Isl., 12 cm

♂ Photo: Ad Konings

A78765-5 *Pseudotropheus (Tropheops)* spec. "Tropheops Chinyankwazi"

Lake Malawi: Chinyankwazi Isl., 12 cm

♀ ♀ Photo: Ad Konings

A78765-5 *Pseudotropheus (Tropheops)* spec. "Tropheops Chinyankwazi"

Lake Malawi: Chinyankwazi Isl., 12 cm

♂ Photo: A.Spreinat

A78760-5 *Pseudotropheus (Tropheops)* spec. "Tropheops Chinyamwezi"

Lake Malawi: Chinyamwezi Isl., 12-13 cm

♂ Photo: A.Spreinat

A78760-5 *Pseudotropheus (Tropheops)* spec. "Tropheops Chinyamwezi"

Lake Malawi: Chinyamwezi Isl., 12-13 cm

♀ Photo: A.Spreinat

A78761-5 *Pseudotropheus (Tropheops)* spec. "Tropheops Chinyamwezi" OB-Morph

Lake Malawi: Chinyamwezi Isl., 12-13 cm

♂ Photo: Ad Konings

A78761-5 *Pseudotropheus (Tropheops)* spec. "Tropheops Chinyamwezi" OB-Morph

Lake Malawi: Chinyamwezi Isl., 12-13 cm

♀ Photo: A.Spreinat

A78820-5 *Pseudotropheus (Tropheops)* spec. "Tropheops Maleri Blue"

Lake Malawi: Nakantenga Isl., 11-12 cm
♂ Photo: Ad Konings

A78820-5 *Pseudotropheus (Tropheops)* spec. "Tropheops Maleri Blue"

Lake Malawi: Nakantenga Isl., 11-12 cm
♀ Photo: Ad Konings

A78815-5 *Pseudotropheus (Tropheops)* spec. "Tropheops Lumessi Blue"

Lake Malawi: Lumessi South (Mozambique), 11-12 cm
♂ Photo: Ad Konings

A78925-5 *Pseudotropheus (Tropheops)* spec. "Tropheops Taiwan"

Lake Malawi: Chisumulu: Taiwan Reef, 12 cm
♂ Photo: A.Spreinat

A78925-5 *Pseudotropheus (Tropheops)* spec. "Tropheops Taiwan"

Lake Malawi: Chisumulu: Taiwan Reef, 12 cm
♂ Photo: Ad Konings

A78925-5 *Pseudotropheus (Tropheops)* spec. "Tropheops Taiwan"

Lake Malawi: Chisumulu: Taiwan Reef, 12 cm
♀ Photo: Ad Konings

A78775-5 *Pseudotropheus (Tropheops)* spec. "Tropheops Chitimba"

Lake Malawi: Chitimba Bay, 10-12 cm
♂ Photo: Ad Konings

A78775-4 *Pseudotropheus (Tropheops)* spec. "Tropheops Chitimba"

Lake Malawi: Chitimba Bay, 10-12 cm
♀ Photo: Ad Konings

A78751-5 *Pseudotropheus (Tropheops)* spec. "Tropheops Chilumba"

Lake Malawi: Mbamba Bay (Tanzania): Ngkuyo Isl., 10-12 cm
♂ Photo: A.Spreinat

A78751-5 *Pseudotropheus (Tropheops)* spec. "Tropheops Chilumba"

Lake Malawi: Mbamba Bay (Tanzania): Ngkuyo Isl., 10-12 cm
♀ Photo: A.Spreinat

A78750-5 *Pseudotropheus (Tropheops)* spec. "Tropheops Chilumba"
Aquarienpopulation
Lake Malawi, 10-12 cm
♂ Photo: E.Schraml

A78750-5 *Pseudotropheus (Tropheops)* spec. "Tropheops Chilumba"
Aquarienpopulation
Lake Malawi, 10-12 cm
♀ Photo: E.Schraml

A78754-5 *Pseudotropheus (Tropheops)* spec. "Tropheops Chilumba"

Lake Malawi: Chilumba: Chitendi Isl., 10-12 cm
♂ Photo: A.Spreinat

A78754-5 *Pseudotropheus (Tropheops)* spec. "Tropheops Chilumba"

Lake Malawi: Chilumba: Chitendi Isl., 10-12 cm
♀ Photo: A.Spreinat

A78740-5 *Pseudotropheus (Tropheops)* spec. "Tropheops Boadzulu"

Lake Malawi: Makokola Reef, 12 cm
♂ Photo: Ad Konings

A78740-5 *Pseudotropheus (Tropheops)* spec. "Tropheops Boadzulu"

Lake Malawi: Makokola Reef, 12 cm
♀ Photo: A.Spreinat

A79240-5 *Pseudotropheus (Tropheops)* spec. "Tropheops Mumbo"

Lake Malawi: Mumbo Isl., 10-11 cm

♂ Photo: Ad Konings

A79240-5 *Pseudotropheus (Tropheops)* spec. "Tropheops Mumbo"

Lake Malawi: Mumbo Isl., 10-11 cm

♀ Photo: A.Spreinat

A79240-5 *Pseudotropheus (Tropheops)* spec. "Tropheops Mumbo"

Lake Malawi: Mumbo Isl., 10-11 cm

♂ Photo: A.Spreinat

A79240-5 *Pseudotropheus (Tropheops)* spec. "Tropheops Mumbo"
Yellow Morph
Lake Malawi: Mumbo Isl., 10-11 cm

♀ Photo: Ad Konings

A78713-5 *Pseudotropheus (Tropheops)* spec.
Albino, Breeding-Form
13 cm

♂ Photo: E.Schraml

A78555-5 *Pseudotropheus (Tropheops) microstoma* Trewavas, 1935

Lake Malawi: Otter Point, 12-13 cm

♂ Photo: Ad Konings

A78555-5 *Pseudotropheus (Tropheops) microstoma* Trewavas, 1935

Lake Malawi: Domwe Isl., 12-13 cm

♂ Photo: W.Deproost

A78555-5 *Pseudotropheus (Tropheops) microstoma* Trewavas, 1935

Lake Malawi: Otter Point, 12-13 cm

♀ Photo: Ad Konings

Schwarm von *Pseudotropheus (Tropheops)* spec. "Tropheops Yellow Head" bei den Tumbi Rocks
School of *Pseudotropheus (Tropheops)* sp. "Tropheops Yellow Head" at Tumbi Rocks

Photo: A.Spreinat

Update Service für Ihre Aqualog Bücher

Update Service for your Aqualog Books

Supplement No.1 to
Loricariidae all l-Numbers
ISBN 3-931702-15-4

Supplement No.2 to
Loricariidae all l-Numbers
ISBN 3-931702-16-2

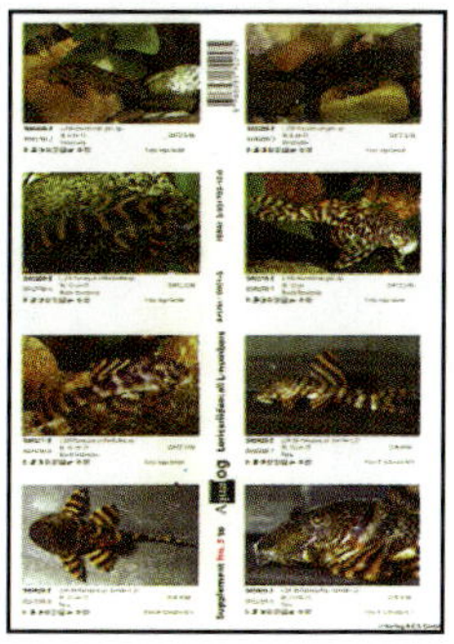

Supplement No.3 to
Loricariidae all l-Numbers
ISBN 3-931702-17-0

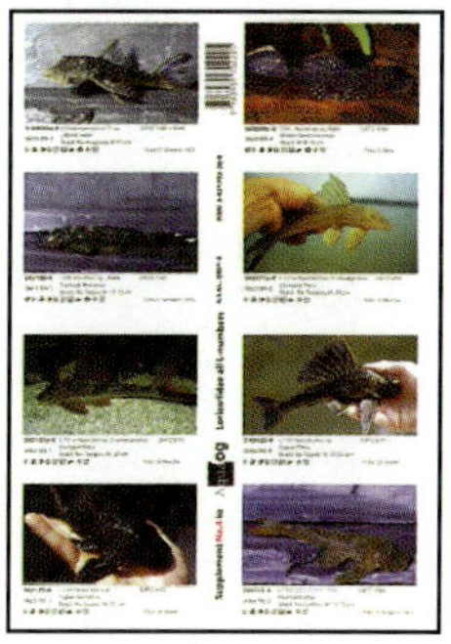

Supplement No.4 to
Loricariidae all l-Numbers
ISBN 3-931702-20-0

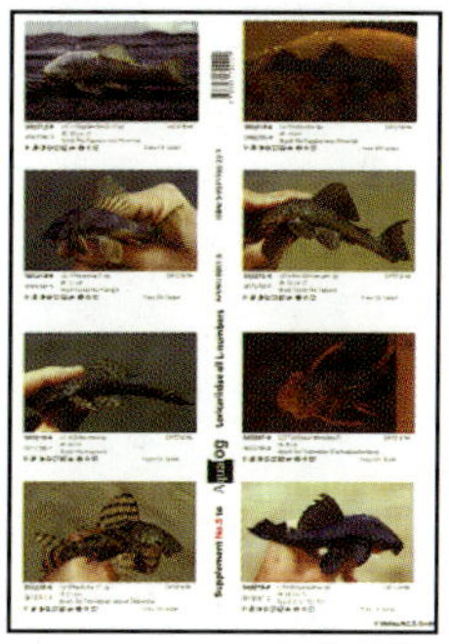

Supplement No.5 to
Loricariidae all l-Numbers
ISBN 3-931702-22-7

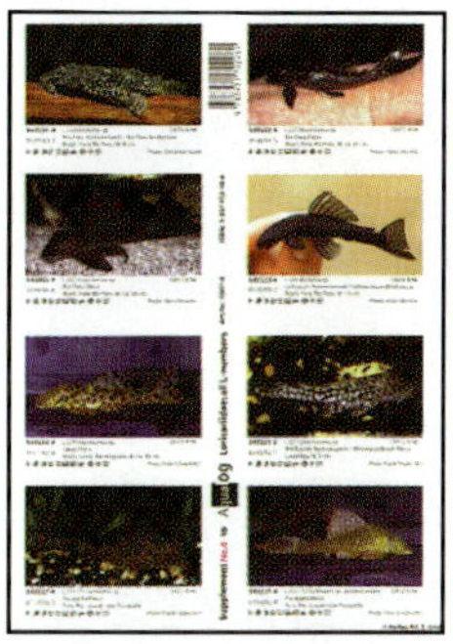

Supplement No.6 to
Loricariidae all l-Numbers
ISBN 3-931702- 28-6

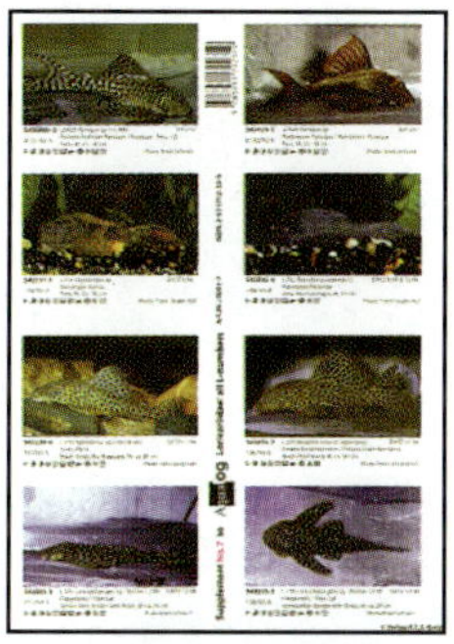

Supplement No.7 to
Loricariidae all l-Numbers
ISBN 3-931702- 35-9

Supplement No.8 to
Loricariidae all l-Numbers
ISBN 3-931702- 72-3

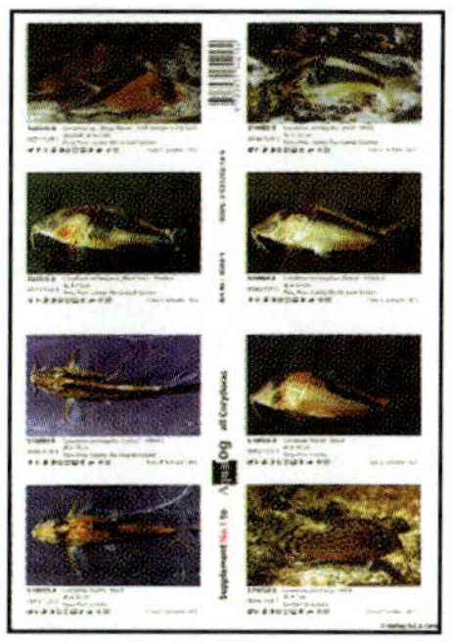

Supplement No.1 to
all Corydoras
ISBN 3-931702-18-9

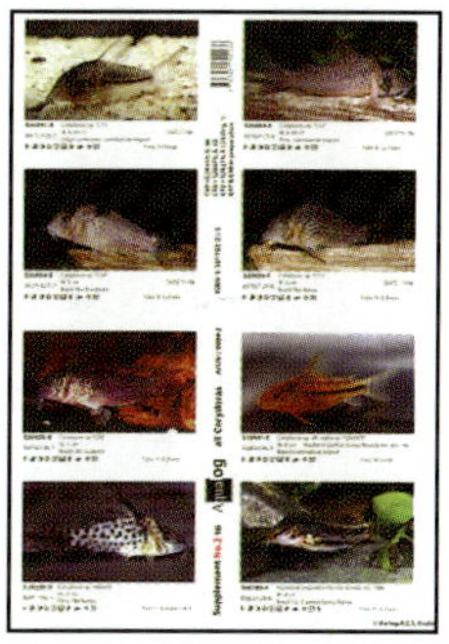

Supplement No.2 to
all Corydoras
ISBN 3-931702-23-5

Supplement No.3 to
all Corydoras
ISBN 3-931702-37-5

Supplement No.4 to
all Corydoras
ISBN 3-931702-83-9

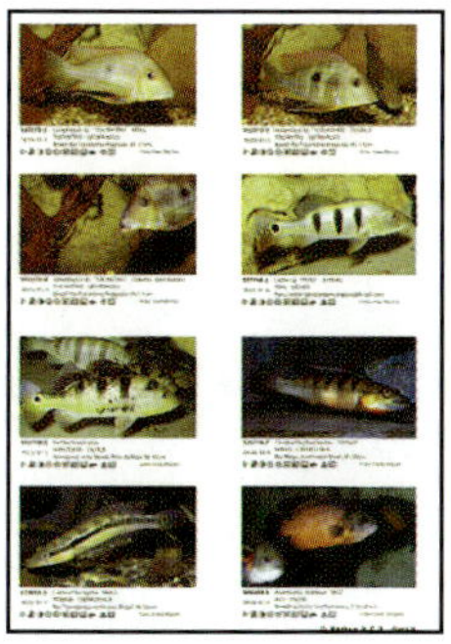

Supplement No.1 to
Southamerican Cichlids 1
ISBN 3-931702-19-2

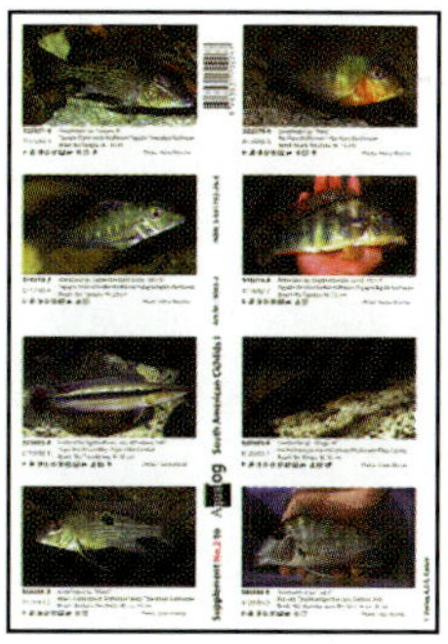

Supplement No.2 to
Southamerican Cichlids 1
ISBN 3-931702-26-X

Supplement No.1 to
Southamerican Cichlids 2
ISBN 3-931702-12-X

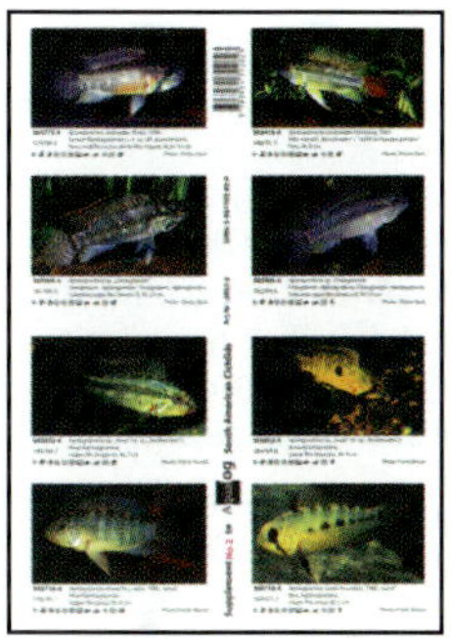

Supplement No.2 to
Southamerican Cichlids 2
ISBN 3-931702-82-0

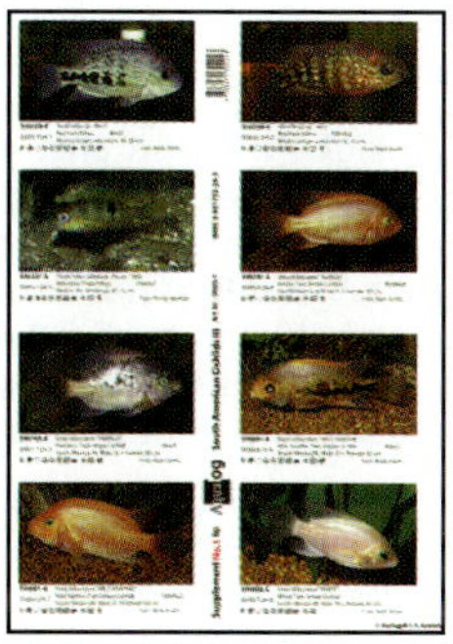

Supplement No.1 to
Southamerican Cichlids 3
ISBN 3-931702-24-3

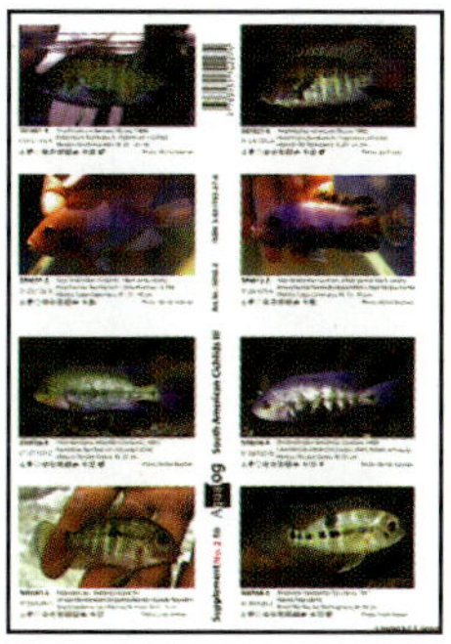

Supplement No.2 to
Southamerican Cichlids 3
ISBN 3-931702- 27-8

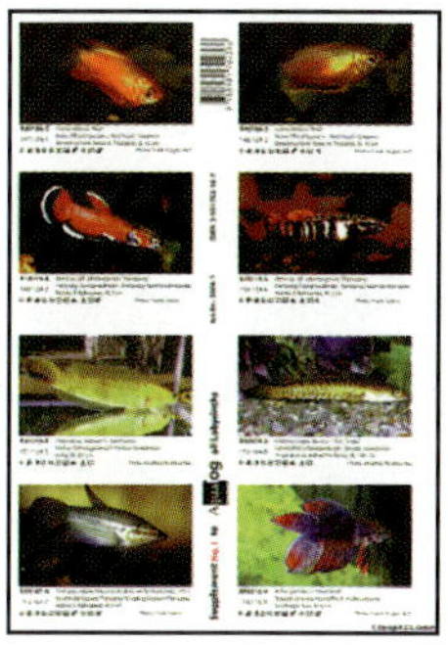

Supplement No.1 to
all Labyrinths
ISBN 3-931702-36-7

Halten Sie Ihren AQUALOG über Jahre aktuell
Keep your AQUALOG up-to-date for years

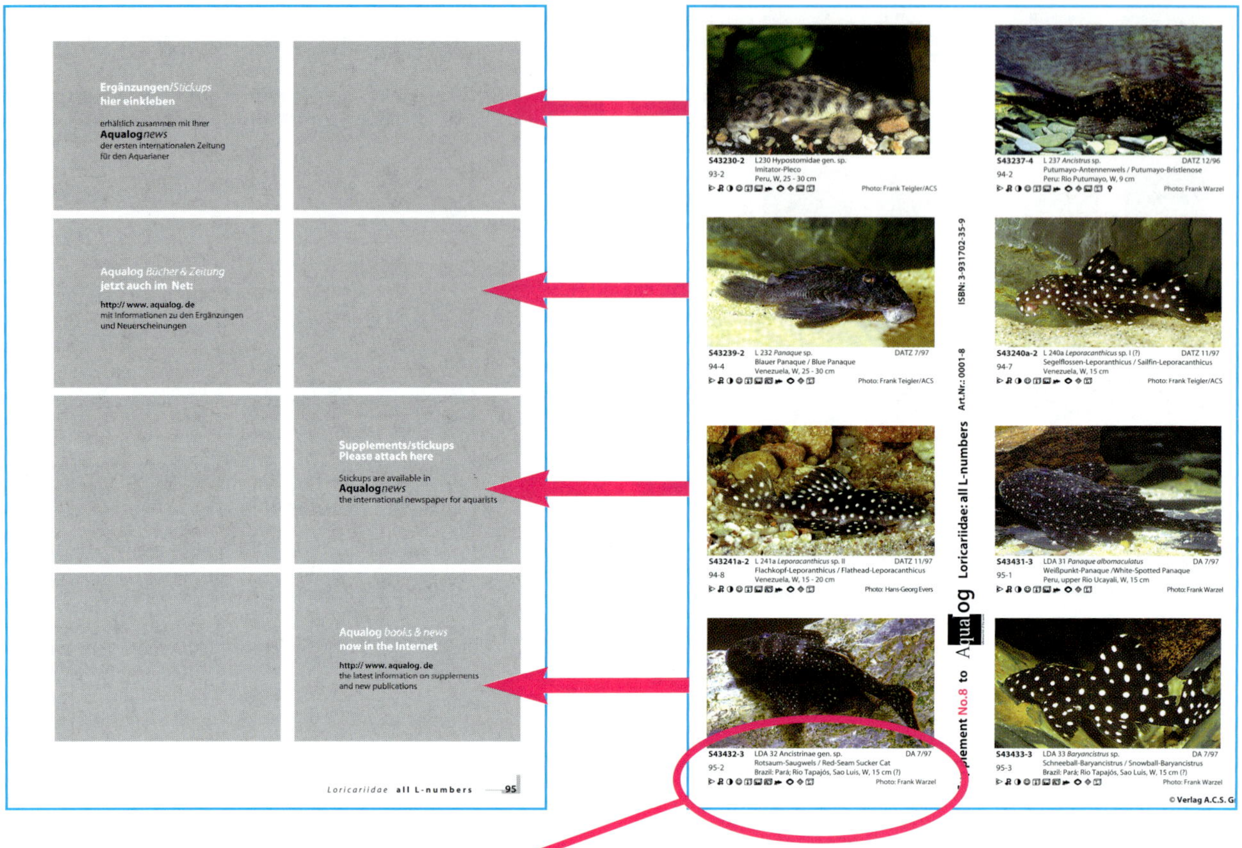

1. **S43432-3** LDA 32 Ancistrinae gen. sp. DA 7/97
2. 170/95-2 Rotsaum-Saugwels / Red-Seam Sucker Cat — Brazil: Pará; Rio Tapajós, Sao Luis, W, 15 cm (?)
3. [Symbol-Leiste]
4. Foto: Frank Warzel

① **Code Nummer**

② **1.Zahl**: fortlaufende Bildnummer.
2.Zahl: Seitennummer des betr. Buches.
3.Zahl: Bildnummer auf der Seite (durchlaufend numeriert von 1-8 von oben links nach unten rechts)

***1.number**: continuous picture-number*
***2.number**: page number in the book*
***3.number:** picture- number on the page (continuously numbered from 1-8 from the top left corner to the down right)*

③ **Symbol Leiste Aqualog-Bücher**
Symbol-text (Aqualog-books)

④ **Bildautor**
Photographer

Die Flutwelle neuer oder neu-importierter Arten reißt nicht ab. Es ist leider unmöglich, sie alle in der Zeitung "AQUALOGnews" als Stickups zu präsentieren. Daher haben wir uns entschlossen, Ergänzungsbögen mit je acht Einklebebildern zu einem Buch herzustellen. Liefer-bar über den guten Zoofachhandel und den Buchhandel zum Preis von DM 4.80 pro Stück. Viel Freude damit! Übrigens: die Stickups aus der news befinden sich nicht nochmals auf den Ergänzungsbögen!

The flood of new or new-imported species doesn´t stop. It is impossible to show them all as stickups in our Newspaper AQUALOGnews. So we decided to print supplements with eight stickers each (each supplement contents pictures for only one volume of AQUALOG). They can be ordered at well-equipped pet-shops or in every bookshop. We hope you enjoy them! By the way: the stickups are not reprinted on the supplements !

Bitte beachten Sie nebenstehendes Schema, bevor Sie die Bilder einkleben. Die Ergänzungen erscheinen nicht zwangsläufig in der Reihenfolge, in der sie eingeklebt werden, sondern in der Reihenfolge ihrer Verfügbarkeit. Wenn wir z.B. anfangs nur das Bild eines Weibchens als Ergänzung haben, jedoch sicher sind, früher oder später auch das Bild eines Männchens zu bekommen, sollte das Bildkästchen links vom Weibchenbild frei bleiben.

Please follow the scheme given here, before you stick in the pic-tures. The supplements are not necessarily in the correct order. For example: if we have only the photo of a female, but we are sure to get the photo of the male sooner or later, too, please keep the space to the left of the female free.

Ergänzungen/*Stickups*
hier einkleben

erhältlich zusammen mit Ihrer
Aqualog*news*
der ersten internationalen Zeitung
für den Aquarianer

Aqualog *Bücher & Zeitung*
jetzt auch im Net:

http:// www. aqualog. de
mit Informationen zu den Ergänzungen
und Neuerscheinungen

Supplements/stickups
Please attach here

Stickups are available in
Aqualog*news*
the international newspaper for aquarists

Aqualog *books & news*
now in the Internet

http:// www. aqualog. de
the latest information on supplements
and new publications

Ihr Nachschlagewerk *your reference work!*

ISBN 3-931702-04-9

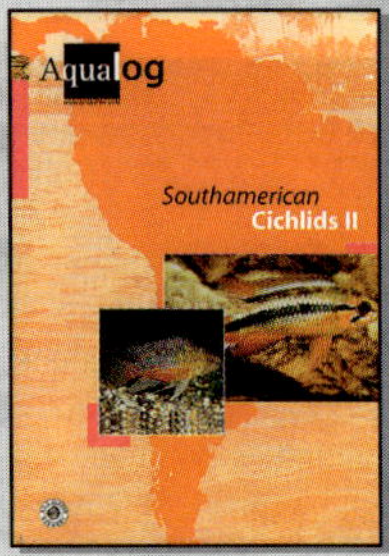

ISBN 3-931702-07-3

ISBN 3-931702-10-3

ISBN 3-931702-01-4

ISBN 3-931702-13-8

ISBN 3-931702-21-9

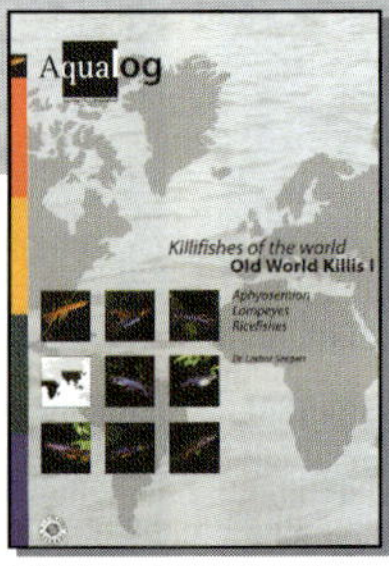

ISBN 3-931702-25-1

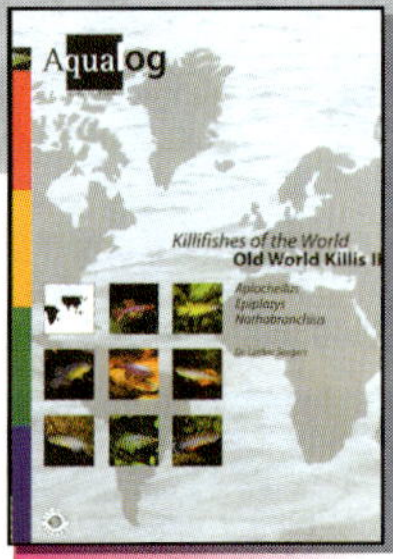

ISBN 3-931702-30-8

Demnächst
coming soon

Killifishes of the World
New World Killis

Autor **Dr. Lothar Seegers**

Demnächst
coming soon

ISBN 3-931702-75-8

ISBN 3-931702-77-4

ISBN 3-931702-78-2

Goldfische und Schleierschwänze
all Goldfish varieties

Autor **Karl-Heinz Bernhardt**

Zu jedem Buch erscheinen auch dekorative Poster!

Full-colour poster to accompany each volume

Vervollständigen Sie Ihr Nachschlagewerk durch weitere Bücher der Aqualog-Reihe! *Complete this reference work with future volumes in the Aqualog series.*

Anfragen an
For more information please contact

Verlag A .C.S. GmbH,
Liebigstraße 1, 63110 Rodgau, Fax: +49 (0)6106 - 644692, email: acs@aqualog.de

Checkliste der nominalen Mbuna-Arten
Checklist of the species and genera of Mbuna

Christyella Trewavas, 1935 (Synonym of *Gephyrochromis)*

Christyella nyasana Trewavas, 1935 (Synonym of *Gephyrochromis moorii)*

Cyathochromis Trewavas, 1935

Cyathochromis obliquidens Trewavas, 1935

Cynotilapia Regan, 1922

Cynotilapia afra (Günther, 1894)
Cynotilapia axelrodi Burgess, 1976

Genyochromis Trewavas, 1935

Genyochromis mento Trewavas, 1935

Gephyrochromis Boulenger, 1901

Gephyrochromis lawsi Fryer, 1957
Gephyrochromis moorii Boulenger, 1901

Iodotropheus Oliver & Loiselle, 1972

Iodotropheus declivitas Stauffer, 1994*
Iodotropheus sprengerae Oliver & Loiselle, 1972
Iodotropheus stuartgranti Konings, 1990

Labeotropheus Ahl, 1926

Labeotropheus curvirostris Ahl, 1926 (Synonym of *L. fuelleborni)*
Labeotropheus fuelleborni Ahl, 1926
Labeotropheus trewavasae Fryer, 1956

Labidochromis Trewavas, 1935

Labidochromis caeruleus Fryer, 1956
Labidochromis chisumulae Lewis, 1982
Labidochromis flavigulis Lewis, 1982
Labidochromis freibergi Johnson, 1974
Labidochromis fryeri Oliver in Davis, 1973 (nom. nud.)
Labidochromis gigas Lewis, 1982
Labidochromis heterodon Lewis, 1982
Labidochromis ianthinus Lewis, 1982
Labidochromis lividus Lewis, 1982
Labidochromis maculicauda Lewis, 1982
Labidochromis margaritae Johnson, 1976 (nom. nud.)
Labidochromis mathotho Burgess & Axelrod, 1976
Labidochromis mbenjii Lewis, 1982
Labidochromis mylodon Lewis, 1982 (status uncertain)
Labidochromis pallidus Lewis, 1982
Labidochromis shiranus Lewis, 1982
Labidochromis strigatus Lewis, 1982
Labidochromis textilis Oliver, 1975
Labidochromis vellicans Trewavas, 1935
Labidochromis zebroides Lewis, 1982

Maylandia Meyer & Foerster, 1984

Maylandia aurora (Burgess, 1976)
Maylandia barlowi (McKaye & Stauffer, 1986)
Maylandia benetos (Stauffer, Bowers, Kellogg, & McKaye, 1997)
Maylandia callainos (Stauffer & Hert, 1992)
Maylandia chrysomallos (Stauffer, Bowers, Kellogg, & McKaye, 1997)
Maylandia cyneusmarginata (Stauffer, Bowers, Kellogg, & McKaye, 1997)
Maylandia elegans (Trewavas, 1935)
Maylandia emmiltos (Stauffer, Bowers, Kellogg, & McKaye, 1997)
Maylandia estherae (Konings, 1995)
Maylandia fainzilberi (Staeck, 1976)
Maylandia greshakei (Meyer & Foerster, 1984)
Maylandia hajomaylandi (in 't Veen, 1984)
Maylandia lanisticola (Burgess, 1976))
Maylandia livingstonii (Boulenger, 1899)
Maylandia lombardoi (Burgess, 1977)
Maylandia mbenjii (Stauffer, Bowers, Kellogg, & McKaye, 1997)
Maylandia melabranchion (Stauffer, Bowers, Kellogg, & McKaye, 1997)
Maylandia phaeos (Stauffer, Bowers, Kellogg, & McKaye, 1997)
Maylandia pursa (Stauffer, 1991)
Maylandia pyrsonotos (Stauffer, Bowers, Kellogg, & McKaye, 1997)
Maylandia sandaracinos (Stauffer, Bowers, Kellogg, & McKaye, 1997)
Maylandia thapsinogen (Stauffer, Bowers, Kellogg, & McKaye, 1997)
Maylandia xanstomachus (Stauffer & Boltz, 1989)
Maylandia zebra (Boulenger, 1899)

Melanochromis Trewavas, 1935

Melanochromis auratus (Boulenger, 1897)
Melanochromis baliodigma Bowers & Stauffer, 1997
Melanochromis benetos Bowers & Stauffer, 1997
Melanochromis brevis Trewavas, 1935
Melanochromis chipokae Johnson, 1975
Melanochromis cyaneorhabdos Bowers & Stauffer, 1997
Melanochromis dialeptos Bowers & Stauffer, 1997
Melanochromis elastodema Bowers & Stauffer, 1997 (Synonym of *M. interruptus)*
Melanochromis exasperatus Burgess, 1976 (Synonym of *M. joanjohnsonae)*
Melanochromis heterochromis Bowers & Stauffer, 1993 (Synonym von *M. vermivorus)*
Melanochromis interruptus Johnson, 1975
Melanochromis joanjohnsonae (Johnson, 1974)
Melanochromis johannii (Eccles, 1973)
Melanochromis labrosus Trewavas, 1935
Melanochromis lepidiadaptes Bowers & Stauffer, 1997
Melanochromis loriae Johnson, 1975
Melanochromis melanopterus Trewavas, 1935
Melanochromis mellitus Johnson, 1976 (Synonym of *M. vermivorus)*
Melanochromis parallelus Burgess & Axelrod, 1976 (?Synonym of *M. chipokae)*
Melanochromis perileucos Bowers & Stauffer, 1997
Melanochromis robustus Johnson, 1986
Melanochromis simulans Eccles, 1973
Melanochromis vermivorus Trewavas, 1935
Melanochromis xanthodigma Bowers & Stauffer, 1997

Metriaclima Stauffer, Bowers, Kellogg, & McKaye, 1997 (Synonym of *Maylandia)*

Microchromis Johnson, 1974

Microchromis zebroides Johnson, 1974

Petrotilapia Trewavas, 1935

Petrotilapia chrysos Stauffer & van Snik, 1996
Petrotilapia genalutea Marsh, 1983
Petrotilapia nigra Marsh, 1983
Petrotilapia tridentiger Trewavas, 1935

Pseudotropheus Regan, 1922

Pseudotropheus (Maylandia) Meyer & Foerster, 1984 (= *Maylandia)*

Pseudotropheus (Pseudotropheus) Regan, 1922

Pseudotropheus (Pseudotropheus) williamsi (Günther, 1894)

Pseudotropheus (Tropheops) Trewavas, 1984

Pseudotropheus (Tropheops) gracilior Trewavas, 1935
Pseudotropheus (Tropheops) macrophthalmus Ahl, 1926
Pseudotropheus (Tropheops) microstoma Trewavas, 1935
Pseudotropheus (Tropheops) novemfasciatus Regan, 1922
Pseudotropheus (Tropheops) tropheops romandi Colombe, 1979 (status uncertain)
Pseudotropheus (Tropheops) tropheops tropheops Regan, 1922

Pseudotropheus elongatus - complex (see Ribbink et. al., 1983)

Pseudotropheus ater Stauffer, 1988
Pseudotropheus cyaneus Stauffer, 1988
Pseudotropheus elongatus Fryer, 1956
Pseudotropheus flavus Stauffer, 1988
Pseudotropheus longior Seegers, 1996
Pseudotropheus modestus Johnson, 1974

Pseudotropheus 'aggressive' - complex (see Ribbink et. al., 1983)

Pseudotropheus fuscoides Fryer, 1956
Pseudotropheus fuscus Trewavas, 1935
Pseudotropheus perspicax (Trewavas, 1935)
Pseudotropheus tursiops Burgess & Axelrod, 1975

Pseudotropheus 'miscellaneous' - complex (see Ribbink et. al., 1983)

Pseudotropheus crabro (Ribbink & Lewis, 1982) (close to *Maylandia)*
Pseudotropheus demasoni Konings, 1994
Pseudotropheus heteropictus Staeck, 1980
Pseudotropheus lucerna Trewavas, 1935
Pseudotropheus minutus Fryer, 1956
Pseudotropheus purpuratus Johnson, 1976
Pseudotropheus saulosi Konings, 1990
Pseudotropheus socolofi Johnson, 1974

INDEX
alphabetisch / alphabetic

INDEX
alphabetisch / alphabetic

Symbols

In order to include as many pictures as possible, and bearing the international nature of the publication in mind, we have intentionally decided against detailed textual descriptions, replacing them by international symbols. This way, one can easily obtain the most important facts about the species and its care.

Continent of origin:

simply check the letter in front of the code-number

A = Africa E = Europe + North America

S = South America X = Asia + Australia

Age:

the last number of the code always stands for the age of the fish in the photo:

1 =	small	(baby, juvenile colouration)
2 =	medium	(young fish / saleable size)
3 =	large	(half-grown / good saleable size)
4 =	XL	(fully grown / adult)
5 =	XXL	(breeder)
6 =	show	(show-fish)

Immediate origin:

W = wild

B = bred

Z = breeding-form

X = cross-breed

Size:

..cm = approximate size these fish can reach as adults.

Sex:

♂ male ♀ female ♂♀ pair

Temperature:

18-22°C	(68 - 72°F)	(room-temperature)
22-25°C	(71 -77°F)	(tropical fish)
24-29°C	(75 - 85°F)	(Discus etc)
10-22°C	(50 - 72°F)	cold

pH-Value:

pH 6,5 - 7,2 no special requirements (neutral)

pH 5,8 - 6,5 prefers soft, slightly acidic water

ph 7,5 - 8,5 prefers hard, alkaline water

Lighting:

bright, plenty of light / sun

not too bright

almost dark

Food:

omnivorous / dry food, no special requirements

food specialist, live food/ frozen food

predator, feed with live fish

plant-eater, supplement with plant food

Swimming:

no special characteristics

in upper area / surface fish

in lower area / floor fish

Aquarium- set up:

only floor and stones etc.

stones / roots / crevices

plant aquarium + stones / roots

Behaviour / reproduction:

keep a pair or a trio

school fish, do not keep less than 10

egg-layer

livebearers / viviparous

mouthbrooder

cavebrooder

bubblenest-builder

algae-eater / glass-cleaner (roots + spinach)

non aggressive fish, easy to keep (mixed aquarium)

difficult to keep, read specialist literature beforehand

warning, extremely difficult, for experienced specialists only

the eggs need a special care

§ protected species (WA), special license required ("CITES")

Minimum tank: capacity:

super small	20 - 40 cm	5 - 20 l
small	40 - 80 cm	40 - 80 l
medium	60 - 100 cm	80 - 200 l
large	100 - 200 cm	200 - 400 l
XL	200 - 400 cm	400 - 3000 l
XXL	over 400 cm	over 3000 l
		(show aquarium)

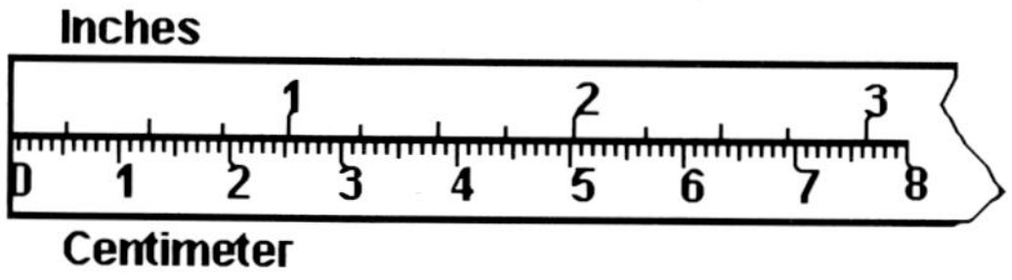